ANIMALS OF THE WORLD EUROPE

Alwyne Wheeler
David Christie
Edwin Cohen
Cathy Jarman
Reg Lanworn

ANIMALS
OF THE WORLD
EUROPE

HAMLYN

Published by
THE HAMLYN
PUBLISHING GROUP LTD
LONDON · NEW YORK · SYDNEY · TORONTO
Hamlyn House, Feltham,
Middlesex, England
© Copyright 1970
The Hamlyn Publishing Group Ltd
SBN 600 10015 4
Phototypeset by Keyspools Ltd
Printed in Hong Kong

CONTENTS

Fishes

Alwyne Wheeler is a professional ichthyologist at the British Museum (Natural History) in London. He is author of several books and scientific papers and is beyond doubt a world authority on the biology of fishes, particularly European species. He has taken part in many field trips and expeditions and is at the moment engaged in revising a standard work on the fishes of British waters.

Birds

David Christie, who wrote the first half of this chapter up to the auks, is a young field ornithologist now living in London and is a member of the London Natural History Society. He has studied birds in their natural environment in many countries of Europe and taken part in important ornithological projects.

Edwin Cohen, who contributed the second half of the chapter, is an active committee member of the International Council for Bird Preservation and has done much valuable work in promoting the conservation of wild birds. He has had a number of papers published and in 1961 a film was made of his garden where almost every bird is colour-ringed and known as an individual.

Mammals

Cathy Jarman is a young zoologist who has recently taken up writing and editing as a career. In the past she has been a popular lecturer at the Zoological Society of London and has written several articles and one complete book on the evolution of animals.

Reptiles

Reg Lanworn is now retired, but he needs no introduction to people who know anything about reptiles. Until recently Mr. Lanworn was overseer of the Reptile House at London Zoo. His television and radio appearances helped to make him well known and these, as well as his field trips to many parts of the world collecting specimens, have given him considerable experience on which to base his writing.

EUROPE

The continent of Europe when discussed as a land mass has some surprising features. For instance it is often described as a peninsula with offshore islands such as Great Britain. Europe is also relatively small when compared with adjoining Asia or Africa. Man has been resident in Europe for a very long time, and mostly because of his efforts to survive and improve his living conditions, the natural fauna and flora has been cut back tremendously. In fact it would be true to say that very little of Europe exists as it does through natural development.

Europe is described in travel brochures and the like as having an equable climate, and this means mild winters and warm but not hot summers. A vegetation map will show clearly that it is in the main a green continent with few desert or dry areas. But most of this land is now cultivated and built on, leaving no room for wild animal populations. Therefore, like most continents, Europe's fauna has been reduced enormously during the last few hundred years during which the human population has been rapidly increasing. Even the seas around the coasts of Europe are heavily fished, to such an extent that certain species have to be protected by limitations on the size of catches or by seasonal restrictions. The species are thus able to reproduce and maintain themselves whereas they might otherwise become extinct.

Until recently no official body cared sufficiently about the protection of wildlife, so that mass cutting back of forest areas and ploughing of farmland, with unlimited methods of pest control, seemed perfectly acceptable. Even the animals and plants of the canals, rivers and seas were not left alone. Effluence from factories and other waste products were poured into the waters, to say nothing of the sewage that added to the pollution.

Fortunately man is now aware of his faults in this sense and has come to terms with nature. Now throughout the whole of Europe organized bodies of people are looking at life through the eyes of future generations and are taking steps to preserve what still exists. In some ways too they are actively putting back part of the wildlife areas that have already been taken, creating wildlife sanctuaries in order to preserve the flora and fauna for the future.

When one sets out to introduce a reader to the animals of a whole continent, one realizes that the difficulty does not lie in finding species to include, but in deciding what to leave out. This has been a very difficult task for the authors. All of them are specialists, all of them feel that their own part of the book should be expanded in order for them to include many more species, even though animals such as insects and amphibians are barely mentioned. Nevertheless the authors have been most co-operative and have provided within this book an interesting and accurate introduction to the animals of Europe which, like others in the series, will form the basis of a better understanding of the world's animals.

FISHES Alwyne Wheeler

Compared with the almost incredible diversity of the fish fauna of tropical regions, such as the fresh waters of Southern America, or the coral reefs of the Indian Ocean, the fishes of Europe seem more or less uniform, rarely so colourful and in general rather dull. To Europeans this is, of course, partly a case of familiarity breeding contempt, for the common elements of the flora or fauna that one grew up with rarely make the same impact that some exotic form does when seen for the first time in later life.

It is true, however, that when compared with a similar north temperate zone such as the fresh waters of North America, Europe is very poorly off for fresh-water fishes. In North America the Great Lakes and St. Lawrence River basin are roughly comparable to the area of north-west Europe (excluding the Danube basin) but the former region contains some 225 native fresh-water fish species while there are fewer than sixty in Europe.

The fishes of Europe should not, however, be dismissed too lightly. Our fresh waters contain fascinating communities of fish adapted to life in habitats as different as the high altitude lakes of the Alps, and the sea-level, salty marshes of the Baltic and the Carmargue. Few of the European fresh-water fishes are found in the British Isles due to the geological history of our offshore islands, but this fact itself is of interest. European sea fish are more varied, but here the contrast between the waters of the Barents Sea and the Mediterranean are so striking that this variation is understandable.

1. To reach its spawning grounds upstream, the Salmon (Salmo salar) *will leap every obstacle. 2. The Cod* (Gadus callarias) *is a most valuable economic fish. 3. The Shanny* (Blennius pholis) *is said to leave the water and bask in the sunshine. 4. Wolf-fish* (Anarhichas spp.) *are cold-water marine fish of the northern hemisphere. 5. The Atlantic Mackerel* (Scomber scombrus) *is a pelagic fish living on small organisms of the upper layers. 6. Pipefishes* (family Syngrathidae), *like sea-horses, have a pouch in which the eggs are hatched.*

One aspect of the European fish fauna makes it of particular interest, and even unique. It is a fauna which has survived in a part of the world where human populations have attained their maximum. Man as a predator has fished in Europe certainly since Upper Palaeolithic times, but the difference between the fish and the mammals which Palaeolithic man hunted is that the mammals are now mostly extinct, or have retreated, but the fish have survived, and indeed are still caught for food in the same area. Man's greatest impact on the fishes of Europe has, however, been through that most insidious of enemies, water pollution. The problem did not emerge, except locally where mining operations discharged lead or other noxious metals into rivers, until the Industrial Revolution. Then, within the last 200 years, the building of factories, mills, paper works, distilleries, all using water and discharging it back into the river in a polluted state, has taken a heavy toll of our fish and caused destruction of their habitats. The damage was not confined to simple pollution, however, for the use of river water in many different roles eventually demanded that the flow of rivers be more evenly controlled; weirs and dams were built, thus preventing migratory fish from freely passing along the river. Within the last thirty years industry has advanced almost to the head waters of our rivers, and giant dams have been built in mountainous regions to provide electrical power for the cities. Thus, as recently as the present century another fish habitat has been invaded, one which hitherto had largely escaped. The heavy concentration of people into big cities throughout Europe has brought its own problems of the supply of drinking water, and the disposal of domestic sewage, the solution (or part-solution) of which have led to further reduction of the habitats available to fish.

But man is still a predator of fish and although he has incidentally damaged the fish stocks in many ways there are still fish to catch, both in fresh water

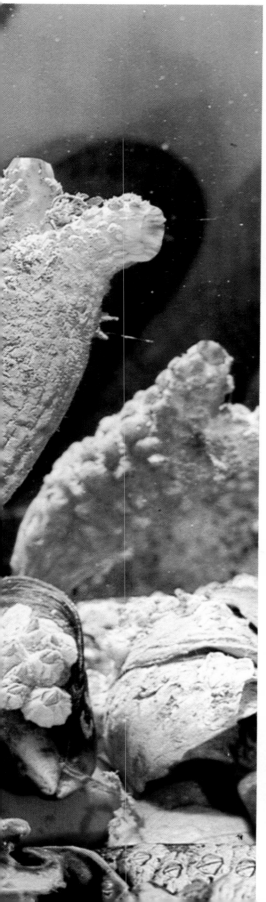

Left: Sea Squirts (Ascidiella aspersa) *with Rock Goby* (Gobius paganellus) *and grey mullet* (Mugil spp.)
Above: Cod (Gadus callarias), *one of the most important European sea fishes.*
Right: the Orange Scorpion Fish (Scorpaena scrofa) *is found in the Mediterranean and eastern Atlantic.*
Below: male Cuckoo Wrasses (Labrus ossifagus) *displaying near the nest. The display stimulates the female, seen behind, to lay eggs on the cleared patch of rock behind the tail of the male.*

and in the sea. Although overfishing is a recent spectre which has risen, and appears to threaten future stocks, man's influence on European fish is not all on the adverse side. Many fish owe their present day abundance to restocking which has bolstered up the declining populations in rivers and lakes. Native fish have been introduced into new waters, often to become established and thrive in them. Exotic species have been brought in from abroad to establish themselves with varying degrees of success, sometimes to become a pest like the American Brown Bullhead (*Ictalurus nebulosus*) and the Black Bullhead (*Ictalurus melas*) in France and Italy, in other places to dwindle in numbers and eventually to disappear. Introductions have been made for so long in Europe that some of the earliest arrivals, such as the carp, are regarded almost as native species today, and the history of their introduction is lost in the mist of time.

Just why are the native fresh-water fishes of Europe so few? The answers lie in the geological history of the northern hemisphere. About a million years ago the climate deteriorated, great masses of ice accumulated in the northern regions and on the mountains further south, forming glaciers which gradually encroached on the lowland areas between. Although the ice retreated at times it returned in four great Ice Ages, during each of which all fresh-water life must have been obliterated. When the ice retreated in the warmer intervals, no doubt some aquatic life returned to the uncovered fresh waters, and we are today living in the aftermath of the fourth glaciation. All the fresh-water fish in northern Europe have entered our waters in the past few thousand years since the ice retreated.

Some fish have entered our rivers from the sea. The salmon, trout, char, whitefishes, and sticklebacks are all capable of living in sea-water, and do so today, although most of them are found commonly in the sea only in northern Europe. There would have been no difficulty in entering the rivers for these species living in the sea near

Above: two members of the carp family, the Common Carp (Cyprinus carpio), *top, and a handsome domesticated variety, the Mirror Carp.*
Right: the elongated, torpedo-shaped body of the Pike (Esox lucius).

the ice limit and generation by generation penetrating northwards as the ice retreated. The majority of Europe's fresh-water fish, however, cannot tolerate a high concentration of salt, and must have migrated slowly westwards from the central Eurasian landmass. Such familiar fish as the Roach (*Rutilus rutilus*), the Chub (*Squalius cephalus*), the Dace (*Leuciscus leuciscus*), the Pike (*Esox lucius*), the Perch (*Perca fluviatilis*) and the two species of bream can only have originated from the east.

The process of recolonizing Europe's fresh waters is not yet complete, as can be seen from the comparative richness of the fish fauna of Germany compared with the British Isles, or on a smaller scale of East Anglian rivers compared with those of the West of England. This gradual east to west impoverishment can be explained by the fact that the fish penetrated westwards slowly following the retreat of the ice. Some passed along the course of the River Rhine into its tributary, the ancient Thames, while these rivers flowed across what is now the bed of the North Sea. Most of the eastern rivers of England were former tributaries of this Thames-Rhine river system, and thus have the same fishes, typical examples of which are the Common Barbel (*Barbus barbus*), the Silver Bream (*Blicca bjoernka*), the Ruffe (*Acerina cernua*) and the Burbot (*Lota lota*). Other fishes either entered the rivers of continental western Europe after the Thames-Rhine connection was broken, or they lived in the headwaters of the rivers, and thus did not reach England. The western English rivers received such true fresh-water fish as they hold either by the 'capture' of tributaries of eastward flowing rivers, and with them their fresh-water fish, or more probably by human agency, either directly by introducing fish, or indirectly by the cutting of canals between river systems.

Ireland forms a special case in that it was cut off from mainland Europe earlier than was Great Britain, and its

last connection was at the north with Scotland, too far north to serve as a bridge for fish to cross. Ireland has relatively few primary fresh-water fish (i.e. those that could not have gained access from the sea), only the Rudd (*Scardinius erythrophthalmus*), the Minnow (*Phoxinus phoxinus*), the Common Bream (*Abramis brama*), the Pike, the Perch and the two species of loach being found there. The Roach and Dace have been introduced within recent historical times. The presence of these seven species, however, does not fit in with the known geological history of Ireland, and some negative evidence exists, notably that there are no genuine Gaelic names for them, and that they are not mentioned in the earliest written records of fish in Ireland, which suggests that in medieval times they too were introduced.

The European fresh-water fish fauna is dominated by two major groups, the salmon family and the carp family. Perhaps best known of the first family is the Atlantic Salmon (*Salmo salar*), a fish which spends the greater part of its life in the sea. The Atlantic Salmon spawns in the early winter, digging a nest for the eggs in the gravel usually well upstream in the rivers. The eggs, which are as large as a good-sized garden pea, drop down between the stones, not to hatch until the spring. Usually they stay in the river for one or two years before passing down towards the sea and the rich feeding it offers. Growth in the sea is remarkable; the smolt which started from its native stream as a two-year-old, six-inch fish may return in eighteen months twenty inches long. On the other hand it may range the North Atlantic for three long winters before returning as a twenty-five-pound summer fish. Within the last few years extensive catches of salmon have been made by commercial fishermen off west Greenland and in the Norwegian Sea. Some of the fish caught have been specimens tagged in the British Isles. The discovery of these oceanic feeding grounds and their exploitation have given rise to the fear that the salmon stocks may be destroyed by overfishing.

Another very familiar member of this dominant family is the trout. In Europe this occurs in two main forms, the Brown Trout, which stays in fresh water all its life, and the Sea Trout, which salmon-like, spends some part of its life in the sea, but both are now recognized as belonging to the same species, *Salmo trutta*. The Brown Trout is a fish which thrives in cold water and it is found in mountain lakes and streams throughout Europe. In these cold waters, however, it rarely grows longer than ten inches, although in the lowland rivers it may grow to two feet or more, and in the large lakes of mainland Europe specimen lake trout of three to four feet and a weight of sixty pounds are known.

The whitefishes share the cold-water mountain lakes with the trout. Unlike that species they are not found in rivers except in the northern parts of their range. Several kinds of whitefish are found in the lowland rivers and lakes of Finland, the U.S.S.R. and Sweden, while others are found in the high lakes in the Swiss and Bavarian mountains, and also in the British Isles. Several lakes in the English Lake District, Lake Bala in North Wales, the Lochs Mabern and Lomond in Scotland and Lough

Neagh in Northern Ireland among others have whitefish in them. Like those populations in mainland Europe, they are thought to be the descendants of fish that entered the lakes in immediately post-glacial times, perhaps of migratory stock that came up rivers from the sea to spawn, and have eventually become adapted to a non-migratory life.

The carp-like fishes (order Ostariophysi) are the other dominant freshwater fish group of Europe. In general they are more tolerant of warmer water and lowered oxygen levels in summertime than are the salmonids. In this way the two groups complement one another; the salmon fishes are dominant in the cooler waters of northern Europe and in high altitudes, the carp fishes are dominant in the lowland rivers and lakes. The division is not absolute of course, as members of both major groups are often found together.

The order Ostariophysi contains many of the most familiar of European fishes; the Roach, Minnow, Bream and Common Carp (*Cyprinus carpio*) being perhaps the best known and most widely distributed.

A less widely distributed relative is the Common Barbel, a fish which although found across Europe south of Denmark, in England is only found in the east from Yorkshire to the Thames. It has been introduced into the River Severn and within recent years it has been released in the River Avon in the west of the country, and in little more than a decade has spread in numbers and range along the river system. This is a relatively large fish, growing to a length of three feet and a weight of up to twelve pounds, and it appears to thrive in relatively deep, swiftly flowing rivers, particularly those with gravelly beds. Its fleshy lips and the long barbels round it mouth show that it is a bottom feeder, and indeed its body shape with a high arched back and flattened belly all indicate a life close to the bed of a river. Several other barbels are found in Europe, the best known of which is the heavily spotted Southern Barbel (*Barbus meridionalis*), known in France as the *Barbeau Truite* for its resemblance to the spotted trout. This species is found in northern Spain, in Italy, southern France, and in a large area of central Europe. It is a smaller species growing to a maximum length of eighteen inches, and is usually found in smaller rivers, and further upstream than the Common Barbel.

The fresh-water breams are represented in the British Isles only by the Silver Bream and the Common Bream. The latter is widely distributed, having even been introduced to Ireland, but the smaller Silver Bream is found only in the rivers of eastern England. In mainland Europe both these fish are widespread north of the Alps, extending northwards to Finland and the Baltic basin, and eastwards far into the U.S.S.R., but in addition there are

Above left: the Minnow (Phoxinus phoxinus) *inhabits clear brooks and streams with a gravel or sandy bottom. Left: although closely related to the Perch, the Ruffe or Pope* (Acerina cernua) *is smaller and not so bright. Right: young Roach* (Rutilus rutilus), *Perch* (Perca fluviatilis), *Three-spined Sticklebacks* (Gasterosteus aculeatus).

several other bream-like fish. The German Zope (*Abramis ballerus*) is found in north Germany and in brackish water around the Baltic. It is most common in deep water, in contrast to its relative, the Zährte (*Vimba vimba*) which is more often found in shallow water in the same area. The Zährte is particularly common in the brackish margins of the Baltic, although it runs up the rivers to spawn in the weedy margins in fresh water. Neither the Zope nor the Zährte grows much longer than twenty inches, but both are caught in small commercial fisheries.

Another typical inhabitant of the edges of the Baltic Sea is the Ziege (*Pelecus cultratus*), which is also known as the Sabre Carp, a stream-lined member of the carp family, with a sharply compressed body giving its belly a knife-like edge. The Ziege is a pelagic, shoaling fish which feeds on small fish such as the Herring and young Whiting. It spawns in fresh water, and like both the Zope and the Zährte is also found in eastern Europe in the rivers running into the Black Sea basin, although it is not found in the region between the two areas.

Some of the smaller carp fishes found in Europe are also of considerable interest on account of their habits. The small minnow-like fish with a brilliant silvery-blue line along its sides, known in Germany as the Moderlieschen (*Leucaspius delineatus*), gets its vernacular name, which means literally 'motherless', from the way the fish appears apparently spontaneously in small newly-dug peat diggings, or quarries. Usually it is an inhabitant of lowland rivers, and is found near the banks, from where it is likely to be swept away during spring floods into such

apparently unpromising waters. It rarely grows longer than three inches. At spawning time in April and May the male has the scattered white tubercules on the head and gill covers that are typical of the carp family. The eggs are looped in strands about the waterplants of the shallows, and the male guards them until they hatch.

Another rather small fish found in central and eastern Europe is the Bitterling (*Rhodeus sericeus*). This too has a silvery, metallic streak along the sides, and the spawning males are beautifully coloured; the normal coloration is heightened in tone, the sides become pearl-like and the fins orange-red. The Bitterling is a fish of slowly flowing rivers, backwaters and lakes, where it is found on sandy or muddy bottoms. At the spawning season the female develops a long egg-laying tube from the vent. The eggs pass through this two at a time, and are deposited in the gill chamber of a swan mussel or a pearl mussel, both of which live partially buried in the bottom mud or sand. The eggs are fertilized when the male ejects a cloud of sperm in the vicinity of the mussel which draws it in through its siphon. The eggs stay within the mussel after hatching, and the young Bitterlings do not leave its shelter until the egg-yolk has been used up and they are about three-quarters of an inch long. By adopting such an unusual spawning place, the Bitterling ensures protection of its eggs and early fry from predators. It also gains a protection from drought, which is a constant threat in the shallow marshy conditions in which it lives, for as the water level recedes the mussel will shift its ground to keep underwater. The advantages of this strange breeding habit are not all on the side of the fish, however, for these fresh-water mussels have an early larval form parasitic on fishes. Bitterling often carry the young of the mussel embedded in their skins.

The loaches are a group of relatives of the carp-like fishes. In Asia they are represented by literally hundreds of known species, but in Europe there are only three species. One of these is the familiar Stone Loach (*Noemacheilus barbatulus*) found very widely in the streams of Great Britain. The Stone Loach is a small fish of some three to four inches, which lives in the smaller streams and rivers and some lakes, usually hiding under stones or in dense weed growth during the daytime. and making feeding forays during the night. The presence of a number of barbels around its mouth suggests that it is a bottom feeder, and most of its food in fact consists of insect larvae which live on the bottom of the river.

The Spined Loach (*Cobitis taenia*) is a small fish flattened from side to side, which appears to burrow in the sand or mud of the slow-flowing waters in which it lives. It is of rather restricted distribution in the British Isles, being found only in eastern England, but eastwards it extends through Europe including southern Scandinavia. In Spain, Italy and the Balkans it is said to be represented by local races or subspecies, and in the Balkans and the Black Sea basin by another closely related species, the Golden Loach (*Cobitis aurata*). Both these species have, as their name suggests, a double pointed spine under each eye, which is usually and swallowing it; the oxygen is removed in a special lining of the intestine and the remaining gas is passed out through the anus. The name Weather Fish is derived from its restless activity in hot weather particularly before thunderstorms; pet fish in aquaria can almost be used as living barometers.

From the rather small loaches to their giant relative, the Catfish or Wels (*Silurus glanis*) may seem a wide step, but they are all members of the order Ostariophysi. Like the loaches, the catfishes are widespread and very numerous in Asia and are found in Africa also, but only one species is found in western Europe. This is a giant of its kind, indeed the largest European freshwater fish, growing to a length of at least fifteen feet and a weight of more than 600 pounds. However, it must be admitted that such very large Wels are

Above left: the Rainbow Trout (Salmo gairdneri irideus), *below, thrives in warmer water than the Brown Trout* (Salmo trutta), *above.*
Above: the Common Eel (Anguilla anguilla) *often travels overland on its journey back to the Atlantic to breed.*
Right: moray eels (family Muraenidae) in Roman amphora off Naples.

retracted within the skin.

The third European loach, the Weather Fish or Pond Loach (*Misgurnus fossilis*) is not found in Britain. It is rather larger than the other species, growing to some ten inches in length. It has the typical elongate shape of the loaches, like them it is well supplied with barbels around the mouth, and is moreover mainly nocturnal in its habits. The Pond Loach lives in stagnant backwaters and ponds and usually lies buried in the muddy bottom. These waters are not infrequently devoid of oxygen for long periods in warm weather. This loach survives in such conditions by gulping air at the surface

more common in the U.S.S.R. than in the Danube and the Baltic basins, which are the westerly limits of their normal range. The Wels has been introduced to the British Isles, and it has lived and bred for some years in the lakes at Woburn Abbey in Bedfordshire.

Apart from its size the Wels is distinguished by the pair of extremely long barbels which with several others border its mouth. It lives in deep water, under the banks or in scoured-out hollows, usually where the bottom is soft. A nocturnal feeder, when young it mainly eats small bottom-living animals, but as it grows larger its diet includes larger items, including waterfowl and small mammals. Legend in central Europe has it that large Wels have been known to eat dogs and even boys.

In contrast to the widely distributed Wels, a second but very localized catfish species is found in south-eastern Europe. This is Aristotle's Catfish (*Silurus aristotelis*), which is found only in the vicinity of the River Achelous in Greece. The breeding habits of this fish were described long ago by the Greek philosopher-naturalist Aristotle, but as his description did not tally with such observations as had been made of the breeding of the Wels in Europe, they were generally discredited. It was not until 1856 when specimens of Greek catfish were examined carefully that it was discovered that Aristotle had been describing another, hitherto unknown, catfish. His observations were vindicated some 2,000 years after his death!

Another large predatory fish living in Europe's lakes and rivers is the Pike. The elongate torpedo-shaped body of the Pike is a familiar sight in summer, as the fish lies at the surface basking in the warmth of the sun. It is widespread across Europe, including the British Isles, and extends continuously across northern Asia and North America (where there are several other species).

Sea-horses (Hippocampus antiquorum) *are feeble swimmers and are carried by currents into brackish and sometimes fresh water.*

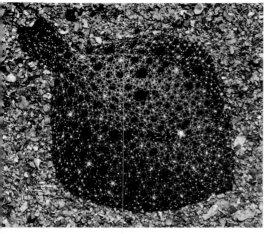

Pike spawn in the spring, usually in the flooded water meadows and along the submerged bankside vegetation, and in such places it is sometimes possible to watch a large female accompanied by two or three smaller males in the act of spawning. Young Pike live in the dense bankside vegetation, and their food is almost entirely composed of the small aquatic invertebrates in this habitat. As it grows, however, it begins to feed on small fish, and as an adult its diet is almost entirely fish. The Pike may grow to a length of forty inches and a weight of around fifty pounds.

is confined to central and eastern Europe. It is well adapted to life in marshy regions, which often suffer a deficiency of oxygen in the warmer months of the year, for it has the ability to hold a bubble of air, gulped at the surface, within its gill chamber.

Some other North American fish have been introduced to Europe, several of them liberated with the object of improving the sport fishing in fresh water. Some of these have been successful, in some cases too successful, for the introduced species has multiplied and spread and become a competitor

Top: Turbot (Rhombus maximus).
Above left: the Thornback Ray (Raja clavata) *is found off European coasts from the Black Sea to the Baltic.*
Above: Lesser Spotted Dogfish (Scyllium canicula).
Left: Plaice (Pleuronectes platessa).

It is a curious contradiction that the closest relative of the Pike is also one of the smallest European fishes. The European Mud-minnow (*Umbra krameri*) grows to a length of four and a half inches, and is a rather short-bodied fish with large scales, brown in general coloration with a bluish tinge along the sides. It lives in the bottom mud of lakes and streams, particularly the flood plains of the lowland rivers, but

of other more valuable native fish.

Not all the introductions of exotic fish have proved harmful, however. The Rainbow Trout (*Salmo gairdneri irideus*) has made a very considerable contribution to fresh-water fisheries, both as a food fish and for sport fishing. Originally a native of north-western America, it is so widely stocked in fish farms and released into rivers and still waters in both Britain and Europe, that it is more common in places than the native trout. In Denmark especially, its culture and stocking have been well developed and it forms an important industry in that country. The Rainbow Trout thrives in rather warmer water than the Brown Trout and its growth rate is also a little faster, and these advantages are turned to good account in lowland areas especially.

A substantial part of the European fresh-water fish fauna is composed of those fish known as secondary freshwater fishes. These are species which could have colonized the rivers by entering them from the sea. The Salmon and the Sea Trout are two well-known examples, the Char is another. A relative of these salmonid fish which has only locally adapted to life in fresh water in places is the Smelt (*Osmerus eperlanus*). It is essentially an estuarine fish which spawns in the middle tidal reaches of large rivers. It seems only to be really common in areas where the salinity is lowered, such as the southern North Sea and in the Baltic, but it occurs in the vicinity of the mouths of most large European rivers, north of Spain.

One of the shads, the Twaite Shad (*Alosa finta*), is also found in landlocked waters in various parts of Europe. The lake of Killarney in southern Ireland is one such, and the Italian lakes Maggiore, Lugano, Como and others all have distinct populations of non-migratory shads. The Twaite Shad looks very like a large herring with a line of black dots along each side, and it belongs to the same family as the Herring, the Sprat and the Pilchard. In addition to the landlocked forms mentioned above, it is relatively common all round the coast of Europe, entering rivers in spring to spawn in the tidal regions well upstream. The Allis Shad (*Alosa alosa*) is larger, and may reach a length of twenty-six inches. It too spawns in the tidal reaches of rivers in the springtime (it is often known as the May-fish from its habit of appearing in that month), but it is much rarer than the Twaite Shad.

The relative scarcity of such fishes, which are typically inhabitants or spawners in the lower reaches of Europe's rivers, is an illustration of the effects of one of the great hazards confronting our fishes—pollution. The industrial and domestic wastes from human populations are largely disposed of by piping them into rivers for carriage into the sea. As the heaviest populations, and thus the biggest cities, are in the low-

Left: the Sea Scorpion (Cottus sp.) *is a member of the bullhead family.*
Above: Common Cuttlefish (Sepia officinalis).
Right: the Lump-sucker (Cyclopterus lumpus) *is often called the Hen-fish.*

lands and often along the lower reaches of the rivers, it follows that it is these areas which are usually the worst affected. Migratory fish, or those that prefer to live in estuaries, have therefore borne the heaviest burden of pollution. The scarcity in the present day of such fish as Salmon in many of the larger rivers such as the Thames, the Rhine and the Seine to name only three is probably as much due to pollution of these rivers as it is due to overfishing and the obstruction of the rivers by building dams and weirs. The Sturgeon (*Acipenser sturio*) in the eastern Atlantic is a very rare fish, indeed one which may be on the verge of local extinction. Its only breeding populations in western Europe are those in Lake Ladoga on the Baltic States (where

distribution of Europe's fishes, and little is known of its biology. Unfortunately there are grave fears that its existence may be threatened by drainage of the marshes in which it lives, and even more that the introduction of an American relative, the Mosquito Fish (*Gambusia*), may exterminate it. *Gambusia* have been released in these marshes to control the numbers of malaria-bearing mosquitos, but they also eat the young of, and compete for food with, Europe's native tooth-carps. Today *Gambusia* can be seen abundantly in the Carmargue, the marshes of Spain and Italy, and on many other parts of the Mediterranean coastline, so while Europe may have gained one fish by this introduction, it may prove to lose two more of its native species.

The Huchen (*Hucho hucho*) a large member of the salmon family, related closely to the chars, was formerly widespread and common in the Danube system; today it is much less common. Overfishing and pollution are held chiefly to blame, but the building of hydro-electric dams in the head waters has further affected it.

it is landlocked), the French river Gironde and the Spanish Guadalquivir. The Sturgeon is occasionally caught along the Atlantic seaboard, but it is not known whether these fish come from these stocks or if they are trans-Atlantic migrants from North American rivers.

The Mediterranean coast of Europe is the only region where members of the family Cyprinodontidae, the tooth-carps, are found in Europe. They are small fishes rarely growing longer than two inches, and are related to the popular tropical aquarium guppies and platies. One species *Aphanius fasciatus* is widespread along the Mediterranean coastline from southern France to Turkey, but two other relatives are very much more restricted; *Aphanius iberus* is found only in eastern Spain and part of northern Africa, while *Valencia hispanica* is found only along the eastern Spanish coastline. The latter is probably the most restricted in

BIRDS David Christie and Edwin Cohen

It has always been difficult to fix a definite eastern boundary to Europe, and the whole of Russia has sometimes been considered as belonging to Asia. However, when introducing the birds of Europe here, we are including western Russia up to the Ural Mountains—a wide expanse of steppe and the breeding grounds of many species—but excluding Turkey. Although Europe is the second smallest of the continents and the most densely populated, it is by no means poor in bird life. From the near-Arctic conditions of Iceland and northern Scandinavia to the comparatively hot climate of southern Spain and the Mediterranean, the variety of birds which may be seen is quite remarkable.

All the four species of divers (family Gaviidae) breed regularly in northern Europe. The Red-throated Diver (*Gavia stellata*) breeds in northern Scotland, Iceland and Scandinavia, and across the Baltic. It has a wider winter distribution than the other divers. A beautiful bird in summer plumage, this diver is easily recognized by its dark crimson throat and upturned bill. Divers are essentially water birds and their legs are set very far back on their bodies; consequently movement on land is impeded and the sight of a bulky diver struggling to reach its nest is quite comical.

Similar in many respects to the divers are the grebes (family Podicipedidae). Probably the best known is the Great Crested Grebe (*Podiceps cristatus*), the largest of the family and quite a magnificent bird in its summer plumage. It builds a floating nest of vegetation and weeds which it anchors to reeds or water plants on the surface of a lake. This grebe has a particularly interesting courtship display in which the two birds face each other on the water and emit a low note which is repeated at increasing speed and ends up as a quacking sound. They then stretch their necks to the fullest extent, rear up with 'tippets' and ear-tufts erected and shake their heads rapidly from side to side as if fencing with their bills. After this, one

The pure white plumage of the Little Egret (Egretta garzetta) *is dazzling.*

or both birds will dive for weeds which are presented to the partner. A much smaller member of the same family is the Little Grebe (*Podiceps ruficollis*) which breeds right across central and southern Europe.

One group of birds which spend almost all their time at sea are the petrels and shearwaters. The Manx Shearwater (*Puffinus puffinus*) breeds in Iceland, the western coasts of the British Isles, north-west France and parts of the Mediterranean. Its nesting colonies may contain huge numbers. Both sexes share the work of excavating a burrow, and the chamber which is to be used as the nest is lined with a few pieces of dry grass or stalks. The one egg is incubated by both birds who take turns at sitting for up to about ten days at a time, going without food until they are relieved by their mate. When the young bird is fully fledged, it is deserted by its parents and remains inside the burrow for two weeks without food before it departs for the first time to begin its life far out at sea. At night the whole colony produces a continuous babbling noise, made up of coos and growls, which may continue throughout the night.

A relative of the Manx Shearwater is the Fulmar (*Fulmaris glacialis*), a completely different looking bird which resembles a stocky gull. It nests in colonies on cliff faces in Iceland, Great Britain and around the coast of central Norway. The Fulmar has the interesting habit of prospecting possible nesting sites, and may regularly visit the same site for a few years before deciding whether to nest there or not. Fulmar numbers have increased greatly in recent years as this bird appears to take great advantage of 'free food' cast

Left: a lone Cormorant (Phalacrocorax carbo). *It lives almost wholly on fish. Above: the Black-throated Diver* (Gavia arctica) *nests on small islands or on the edge of a lake.*

overboard by fishing fleets in the North Atlantic.

The storm petrels (family Hydrobatidae), which are smaller and blackish in colour with white rumps, lead the same sort of life as the shearwaters.

Belonging to a different order of birds but also a maritime species is the Gannet (*Sula bassana*). This bird breeds in Iceland and on some of the islands off the west coast of Scotland and south coast of Ireland. There is one colony off south-west Wales, and Gannets also breed in Brittany, the Channel Islands, Norway and the islands between Iceland and Scotland. Bass Rock in the Shetland Isles is now famous for its huge colony of Gannets which in 1966 numbered as many as 6,000 pairs. The Gannet is a spectacular sight when it is feeding. From a height it dives headlong, wings half closed, and enters the water like a bullet with a loud splash, closing its wings just in time.

Pelicans (family Pelecanidae) and cormorants (family Phalacrocoracidae) are also fish-eaters. The Dalmation Pelican (*Pelecanus crispus*) is a huge bird with a wing span of eight or more feet. It nests in colonies among reeds in coastal lagoons and marshes and inland lakes, where it catches fish from the surface as it swims along. Its breeding range is confined to the delta of the Danube, the south-east Adriatic and north and south of the Caspian Sea.

More widespread than either the pelicans or Gannet is the Cormorant (*Phalacrocorax carbo*) and its close

The wing span of the Gannet (Sula bassana) *reaches 6 feet. It takes 5 years to attain adult plumage.*

relative the Shag (*Phalacrocorax aristotelis*). Both these are large, dark birds, living almost entirely on fish which they catch by diving from the surface. The Cormorant breeds on rocky ledges along the northern coasts of Europe and in the south-east among trees by inland lakes.

Nesting in trees and reeds over a large part of Europe are the herons (family Ardeidae). The Bittern (*Botaurus stellaris*) nests over much of central and southern Europe across to eastern Siberia and also locally in England and Sweden. It is a large golden-brown bird with black markings which make an ideal camouflage, for it lives among dense reed-beds, and when alarmed it points its neck and bill straight upwards as it stands motionless, and so is difficult to distinguish from the surrounding reeds. In spring the male makes a resonant booming sound which in still weather carries for miles.

The most common member of the family, and also the largest, is the Grey Heron (*Ardea cinerea*) which is found on the Norwegian coast and right across central and southern Europe. It breeds in Norway, Great Britain and the whole of central Europe. Colonies of nests, called heronries, are built at the tops of trees.

Almost as large is the Great White Heron (*Egretta alba*) whose numbers have sadly been much diminished during the last century by slaughter, the long white feathers having been very popular for ladies' hats. At present it breeds in southern Russia and on the Danube in Hungary, in dense reed-beds.

A bird known to many is the Stork, correctly called the White Stork (*Ciconia ciconia*). The Stork breeds from the Baltic countries south to the eastern Mediterranean and in Spain. Although it will build its huge nest of sticks in treetops, it is best known for its habit of nesting on roofs of houses in towns and villages. It can sometimes be a great nuisance and in many places special poles are erected to attract these birds away from houses. In Germany, the numbers of Storks have been fewer in recent years, a possible cause being the increased use of insecticides which poison the food of this and of many other bird species.

The 'national bird' of the Netherlands is the Spoonbill (*Platalea leucorodia*), a summer visitor to shallow waters and estuaries in Holland, southern Spain and south and east of Austria across to the Caspian Sea. Its long spatulate bill gives it both its common and scientific names.

The Flamingo (*Phoenicopterus ruber*) is resident in southern France, in the Camargue where it breeds, and in southern Spain where it may sometimes breed. An unmistakable bird with its pinky-white plumage and crimson and black wings, it has a most peculiar downward-bent bill which it moves upside-down and from side to side on the bottom of shallow lakes, retaining food and sifting out water and mud.

One of the most interesting and attractive groups of birds are the ducks, geese and swans. There are three species of swan, the most familiar being the Mute Swan (*Cygnus olor*) which is found in many places in a semi-domesticated state. However, the most striking of Europe's swans is surely the Whooper Swan (*Cygnus cygnus*) which is distinguished by the large triangular yellow patch on the side of its bill, and by its neck which is held straight and erect. It breeds only in Iceland, northern Scandinavia and Russia.

The largest of the European grey geese and also the most common is the

Bean Goose (*Anser fabalis*), its head and neck contrasting with the paler breast, but the most beautiful of the geese to be seen in Europe is the very small Red-breasted Goose (*Branta ruficollis*). Its striking black, white and chestnut plumage renders it unmistakable if the bird watcher is lucky enough to see it. It breeds in north-east Russia but does not cross to the European side of the Ural Mountains except on rare occasions in winter. However, a small number do spend the winter months in Hungary.

One of the best known of all European birds is the Mallard (*Anas platyrhynchos*), the ancestor of the domestic duck. This large duck is found throughout Europe and is commonly seen on lakes in parks and on rivers where it is only too willing to accept titbits from people. The drake's glossy green head and dark chestnut breast, separated by a narrow white collar, are characteristic, as are the blue secondary feathers on the wing which are called the speculum. The mottled brown female is by comparison a drab bird. After the breeding season, from about June or July, drakes of every duck species undergo an annual

moult of their normally bright plumage and become very similar in appearance to the females. This stage is called the eclipse and lasts until autumn.

The smallest of European ducks is the attractive Teal (*Anas crecca*), which is no more than about fourteen inches in length. It is almost as widespread as the previous species. A duck more characteristic of the north is the Wigeon (*Anas penelope*) which breeds in Iceland and Scotland and across the north of Europe, extending as far south as Germany. Its whistling call 'whee-oo', often heard issuing from a bank of mist at dawn, gives away its presence when the bird cannot be seen. In winter huge numbers gather in estuaries and on mud-flats, and many winter on large inland lakes.

Left: 3 waders frequently seen on passage: from left—Turnstone (Arenaria interpres), *Knot* (Calidris canutus) *and Sanderling* (Calidris alba).
Above: Common Guillemots (Uria aalge) *breed on narrow cliff ledges.*

These are all surface-feeding ducks, or dabbling ducks as they are often called, and are likely to be seen inland or not far from land. The diving ducks may well be seen out at sea although most of them are also regular inhabitants of large inland waters. One of the most beautiful is the Red-crested Pochard (*Netta rufina*), the male of which has a fiery red head with a bright red bill. It breeds in the Mediterranean region and southern Russia and also in Holland, Denmark and Germany, where in recent years its breeding areas have extended and its numbers increased. It is often kept in captivity and not infrequently some of these birds escape. Another bird also kept widely in captivity is the Tufted Duck (*Aythya fuligula*). Its distribution in Europe is far more widespread and it breeds right across the north of the continent. The drake is black and white and its drooping crest is visible from some distance away.

The largest of the sea ducks is the Eider (*Somateria mollissima*), both sexes of which have peculiar long sloping bills. It is mainly resident in its breeding areas, Iceland and the northern coasts of Europe, although in recent years it has begun to move further south in winter. The Eider's nest is composed of grass and seaweed, copiously lined with feathers and down from the duck's breast. This is collected for use in eiderdowns and pillows, but as soon as the down is taken from her nest the female immediately replaces it with more.

Two species of scoter breed in the extreme north. The most numerous of these is the Common Scoter (*Melanitta nigra*), the drake of which is completely black except for a conspicuous yellow patch on the bill which also has a large knob at the base. Scoters are essentially sea ducks, coming ashore normally only to breed. However, in winter odd birds are occasionally seen on inland lakes, and after gales may sometimes

be stranded along the shoreline. Another essentially maritime species is the Long-tailed Duck (*Clangula hyemalis*) which nests in Iceland and northern Scandinavia. This is one of the noisiest ducks, the male uttering a four-syllable, nasal note. When heard from a flock at sea, the effect is strangely musical.

The sawbills are a group of diving ducks which have a serrated edge to their bill, used to hold fish. Three are seen regularly in Europe, the commonest being the Red-breasted Merganser (*Mergus serrator*), an attractive duck with a chestnut breast band and greenish-black head with a double crest. The female is dull brown-grey above with a chestnut head.

An entirely different order of birds is the Falconiformes, the diurnal birds of prey. The smallest of the vultures is the Egyptian Vulture (*Neophron percnopterus*) which is also the most common. It often visits rubbish dumps in villages or towns and sometimes a small group will join other vultures and devour what they have left. It is resident in Spain and southern Europe. A bird of mountainous country in parts of the south is the Griffon Vulture (*Gyps fulvus*) which is much larger and in flight shows very long wings with primaries widely spread. It breeds sociably on cliffs, and hundreds gather in spring on the cliffs of southern Spain.

One of the best-known birds of prey is the magnificent Golden Eagle (*Aquila chrysaetus*) which may be almost three feet in length. It is recognized by its majestic flight, soaring and gliding with occasional wing beats. This is the most widespread of the eagles, although sadly it is becoming increasingly rarer along with so many other birds of prey. It is resident in Spain and the Mediterranean countries, Scotland, Norway and across central Russia. Sometimes the same nest is used for a number of consecutive years and extra branches are added each year, making the nest a huge structure.

Above: Flamingo (*Phoenicopterus ruber*).
Left: Puffin (*Fratercula arctica*) *with its bill full of sand eels.*

The Imperial Eagle (*Aquila heliaca*) is similar but has white feathers on the scapulars. It is restricted to the south-east of Europe and to Spain. The Spanish race has pure white on the shoulders as well as the scapulars. The much more common and widespread Buzzard (*Buteo buteo*) breeds in most parts of Europe apart from the extreme north. Over much of this area it is a resident. The plumage is remarkably variable but generally is dark brown with paler mottling on the underparts. It often soars at a great height on broad, rounded wings, sometimes unnoticed apart from the plaintive mewing cry which it utters. Similar to this species are the rarer Rough-legged Buzzard (*Buteo lagopus*) and Honey Buzzard (*Pernis apivorus*), both of which are more restricted in range.

The huge White-tailed Eagle (*Haliaeetus albicilla*) frequents rocky coasts and inland lakes in north and east Europe; it catches fish, birds and mammals, even as large as a young deer.

Once a familiar bird in Europe, the Sparrow Hawk (*Accipiter nisus*) has recently become rare in many places although there are signs here and there of a recovery. This is a bird of the woodlands and farms which hunts small birds flying alongside hedgerows, slipping from one side to the other and suddenly pouncing. The male is much smaller than the female.

Resembling the female Sparrow Hawk but much larger is the Goshawk (*Accipiter gentilis*) which hunts for prey among trees, flying fast and low.

Three species of kite breed in Europe. The two best known are the Red Kite (*Milvus milvus*) and the Black Kite (*Milvus migrans*) both of which breed in south-west and central Europe and the Mediterranean countries. The Red Kite also breeds in Wales (although it is estimated that there are very few pairs left there now, despite rigid protection) and the Black Kite breeds across most of Russia. The Black Kite often feeds on carrion and may even be seen scavenging around refuse dumps.

The Osprey (*Pandion haliaetus*) is the sole member of its family (Pandionidae).

Left: a pair of White Storks (Ciconia ciconia) *at their nest in Greece.*
Above: an imposing view of a pair of graceful Mute Swans (Cygnus olor) *in flight, their necks outstretched.*

In Europe it breeds in the east from Denmark to Russia and in parts of the Mediterranean. Recently it has also returned to Scotland to breed, but in 1963 when a pair failed to hatch a single egg an analysis revealed that quantities of the pesticides DDT, aldrin and dieldrin were present in the eggs. The Osprey lives almost entirely on fish, which shows that these pesticides are even finding their way to the sea and polluting the life there. The Osprey's method of catching prey is remarkable —immediately it spies a fish below, it hovers at a height for a moment before plunging feet first into the water and grasping the fish in its strong talons.

Of the four species of harrier found in Europe the Pallid Harrier (*Circus macrourus*) is confined to Russia and the south-east. Montagu's Harrier (*Circus pygargus*), the Hen Harrier (*Circus cyaneus*) and the Marsh Harrier (*Circus aeruginosus*) are distributed over most of the continent apart from the north, although the Hen Harrier is absent from the south in summer but breeds also in Scandinavia and northern Britain. In winter Montagu's Harrier migrates to Africa, the Hen Harrier moves into west and south Europe and the Marsh Harrier to southern Europe and Africa. Harriers have a characteristic flight, gliding low with wings held in a shallow 'V'. The Marsh Harrier hunts over dense reed-beds and drops into the reeds to catch its prey.

Falcons are fast, streamlined birds of prey with long pointed wings. The most common is the Kestrel (*Falco tinnunculus*) which even nests on buildings in large towns and cities. In Belgium, where birds of prey receive virtually no legal protection, the Belgian section of the International Council for Bird Preservation has elected this bird as its 'national bird' to draw attention to the need for conservation. The most characteristic mark of this species is its habit of hovering for minutes at a time some thirty feet up in the air before moving away a short distance only to start hovering again as it surveys the

ground for food. The Hobby (*Falco subbuteo*), on the other hand, is a falcon of dashing flight, relying on its speed and agility to catch small birds and insects in the air.

In eastern Europe a small summer visitor is the Red-footed Falcon (*Falco vespertinus*), the male of which is very dark grey with chestnut under-tail coverts and a bright red bill and legs. On the rocky islands and cliffs of the Mediterranean, Eleonora's Falcon (*Falco eleonorae*) breeds. The favoured places seem to be the Greek islands, and there are several colonies on the cliffs of Cyprus.

The fastest falcons are the Peregrine (*Falco peregrinus*) and the Gyrfalcon (*Falco rusticolus*). The Peregrine is absent from Iceland but breeds over most of the rest of Europe, although its numbers have decreased alarmingly in the last ten years, a chief cause being the increased use of agricultural chemicals. It is an expert in the air, catching birds by stooping on them at incredible speed (up to 200 miles per hour in some cases) and striking them with great force to the ground.

Above right: the Common Gull (Larus canus) *is in fact one of the least common of gulls in many parts of Europe.*
Below right: the only phalarope to breed in Europe is the small Red-necked Phalarope (Phalaropus lobatus).

In wild open country and sea coasts in Norway and Arctic Europe lives the largest of the European falcons, the Gyrfalcon. The Iceland Falcon (*Falco rusticolus islandus*) is slightly whiter on the mantle than the Scandinavian race, and the Greenland Falcon (*Falco rusticolus candicans*), a rare visitor to Britain, is almost pure white. The Iceland section of the International Council for Bird Preservation has adopted the race which breeds there as its 'national bird' in the hope that the necessary interest in conservation will be stimulated.

The order Galliformes comprises the grouse (family Tetraonidae), the partridges and pheasants (Phasianidae) and the rare Andalusian Hemipode (*Turnix sylvatica*) which breeds very locally in Spain and Portugal. In northern Europe the Willow Grouse (*Lagopus lagopus*) breeds on moors and is a resident. The

Left: Gannet (Sula bassana) *with young on Bass Rock.*
Above: the Purple Heron (Ardea purpurea) *is a summer visitor to parts of south and east Europe.*
Below: the typical nest site of the Grey Heron (Ardea cinerea) *is the top of a tall tree.*
Right: Puffins (Fratercula arctica) *spend much of their time staring out to sea.*

*Above: most common European marsh bird is the Redshank (*Tringa totanus*). Left: the Crane (*Grus grus*) is normally a very shy bird. In spring it performs a remarkable display dance.*

cock bird is dark brown with white wings and in winter both sexes are pure white with a black tail. When flushed, this species has a habit of looking back over its wing as it flies away. The British race, the Red Grouse (*Lagopus lagopus scoticus*) lacks the white wings and white plumage in winter.

The Black Grouse (*Lyrurus tetrix*) is more widespread, breeding as far south as the Balkans and across Russia. The bluish-black plumage and lyre-shaped tail of the male (Blackcock) are distinctive, while the female (Greyhen) is grey-brown. In spring the birds assemble at their display-grounds, called 'leks', where the cocks fight over the hens. The plumage is puffed out, the wings droop and the tail is vertically raised in display.

The largest of the grouse is the huge Capercaillie (*Tetrao urogallus*) of the coniferous forests of north and central Europe, Russia and the Pyrenees. The male is almost three feet in length and is blackish with a fan-shaped tail.

Across most of central and parts of northern and southern Europe, the Partridge (*Perdix perdix*) is found on farmland and pasture-land where it nests under hedges and bushes or in grass, laying up to twenty eggs. In winter family flocks of these birds may be seen feeding together, although in recent years numbers have diminished, particularly in Britain, and there is some concern that the species may be slowly disappearing from some parts of the continent. The Red-legged Partridge (*Alectoris rufa*) of the Iberian Peninsula, France and England is resident in drier parts of those countries.

Well known to sportsmen everywhere is the Pheasant (*Phasianus colchicus*) which since its introduction to Europe in Roman times has spread to most parts of Great Britain and central Europe. Many are bred each year on game farms for introduction into the farmlands and woodlands which are its favourite habitat.

The Crane (*Grus grus*) is a large bird (forty-five inches long) with an upright stance. It has very long secondary feathers on the wings which form what looks like a black, drooping tail. On its breeding grounds—in Scandinavia

Above: a solitary Curlew (Numenius arquata) *on the Dee Estuary, Cheshire.*
Above right: the most common duck in Europe is the Mallard (Anas platyrhynchos).
Right: the Bittern (Botaurus stellaris) *adopting the 'neck-up' posture seems to merge into its background.*

and Germany eastwards to the Volga, and parts of the south-east of Europe—it performs a remarkable display dance in spring, when it bows and jumps in a very picturesque ceremony. In winter it moves south, impressive flocks migrating in a long line or 'V' formation.

Rails, crakes and coots (family Rallidae) are marsh birds, most of them not frequently seen as they prefer to skulk among reeds. The Water Rail (*Rallus aquaticus*) is the only one found in Iceland. It is distributed over most of western Europe from Greece and is also resident in the area of Transcaucasia. It spends almost all its time in reed-beds searching for food, and generally gives away its presence by its voice, heard more at night, which is a very characteristic assembly of grunts, screams and squeals.

Two familiar birds on ponds and lakes are the Moorhen (*Gallinula chloropus*) and Coot (*Fulica atra*). The

Above: the unmistakable Kingfisher (Alcedo atthis) *is a brilliant sight as it flashes past.*
Right: the Coot (Fulica atra) *builds its nest among reeds on a lake.*

Moorhen is common on small farmyard ponds, whereas the Coot prefers larger lakes and rivers, although it will often visit small expanses of water. The Coot is black all over with a conspicuous white frontal shield. When taking flight, it patters along the surface of the water for some distance before taking off.

Bustards (family Otididae) are shy birds of grassy steppes and cultivated land. The Great Bustard (*Otis tarda*) is a very large bird with a grey neck and sandy upperparts barred with black. It is resident in Spain, East Germany, the Balkans, and south Russia, nesting on the ground.

The large order Charadriiformes includes the waders, gulls, skuas, terns and auks. The Oystercatcher (*Haematopus ostralegus*), essentially a bird of the shore, has an interesting display. A small group will assemble and perform curious 'piping' ceremonies; with quick steps they dance about excitedly, uttering piping trills which end in a duet between male and female.

Widespread in most parts of Europe is the Lapwing (*Vanellus vanellus*), a familiar bird on farmland and a friend of the farmer as it eats many insects, worms and spiders. It has broad, rounded wings and the wing beats are slow and flapping. The call-note is a loud 'peer-weet'. When displaying the male rises quite high in the air and then plunges down, sharply turning just before it reaches the ground, and flies away, twisting and turning, to repeat the spectacle.

Nesting among heather in Iceland and the north of the continent, is the Golden Plover (*Pluvialis apricaria*). In summer it has golden upperparts and is black below with a broad white band from the forehead to the flanks. Along the western shores of Europe a small bird with a sandy-brown back and broad black breastband may often be

Above left: Great Skua (Stercorarius skua) *in an aggressive posture.*
Left: the Peregrine Falcon (Falco peregrinus) *usually nests on a steep cliff or crag of a mountain.*
Right: a Kestrel (Falco tinunculus) *braking before landing.*

Above: the beautiful male Pheasant (Phasianus colchicus) *is highly prized as a game bird.*
Above right: the nest of a Swallow (Hirundo rustica), *made of mud and straw and built on a ledge in a barn.*
Right: the Griffon Vulture (Gyps fulvus) *inhabits mountainous country in parts of southern Europe.*
Opposite page left: the Hooded Crow (Corvus corone cornix) *is a sub-species of the Carrion Crow.*
Far right: the Middle-spotted Woodpecker (Dendrocopos medius) *is confined to central and south-east Europe.*

seen in winter running about and stopping briefly to pick up food. This is the Ringed Plover (*Charadrius hiaticula*) which breeds in Europe only in the north.

The Snipe (*Gallinago gallinago*) prefers swamps and bogs and is rarely seen along the shore-line. It is a plump bird with a long bill and is usually seen only when flushed, often by accident, when its harsh cry as it zig-zags away is characteristic. This species has a peculiar drumming flight, most frequently from about March to June, in which it dives at an angle of forty-five degrees with its tail feathers spread out. The outer tail feathers vibrate causing a bleating, tremulous sound. The similar Woodcock (*Scolopax rusticola*) is a woodland bird and is nocturnal in habits. Its thickly barred russet-brown plumage is an ideal camouflage which, with its secretive habits, makes it difficult to observe.

Left: Golden Eagle (*Aquila chrysaetus*).
Below: gathered around a carcase in southern Europe, Egyptian Vultures (Neophron percnopterus), *a Raven* (Corvus corax) *and a Black Kite* (Milvus migrans).
Right: Osprey (Pandion haliaetus).

Of the many waders seen in Europe the Dunlin (*Calidris alpina*) is the most common. It is generally only seen on or near the coast, often in immense flocks numbering many thousands, and it nests on northern marshes and moors.

The Greenshank (*Tringa nebularia*) has a long, slightly upturned bill, long green legs and a white rump and lower back. Its characteristic note 'tew-tew-tew' is usually repeated as it rises from a secluded corner of a marsh. Its breeding grounds are the moors and forests of Scotland and Norway across to Russia.

The largest European wader is the Curlew (*Numenius arquata*), a widespread breeding bird over much of the continent. Its long down-curved bill is unmistakable as is its call and the incessant bubbling sound produced by a flock feeding on mud-flats. Often seen with Curlews is the Black-tailed Godwit (*Limosa limosa*) which in summer is rich chestnut on the neck and breast.

Phalaropes (family Phalaropidae) are mostly oceanic during the winter but come inland to breed. The Red-necked Phalarope (*Phalaropus lobatus*) is a small northern bird with a very fine bill.

display grounds where the polygamous males fight while the females (called Reeves) look on.

On the coasts of the Baltic and North Seas, a small area in central Europe, and the Mediterranean and Black Seas, the Avocet (*Recurvirostra avosetta*) breeds locally. This striking black and white bird has a long uplifted bill which it skims from side to side in shallow water in search of food. Its legs are very long but in comparison with those of the Black-winged Stilt (*Himantopus himantopus*) are short, for the latter bird appears to have legs far too long for the

It swims round and round on the water picking up insects and is exceedingly tame. Like all phalaropes the female is more brightly coloured than the male and in spring and summer is quite beautiful with a bright orange patch down each side of the neck and with white throat and underparts.

An interesting and unusual wader is the Ruff (*Philomachus pugnax*). The male in breeding plumage has a remarkable ruff and ear-tufts which vary in colour depending on the individual and may be black, chestnut, purplish-buff or white or a combination of any of these colours. On marshes and meadows in their breeding areas from Holland eastwards, the birds gather on

rest of its body. Its black and white plumage combined with these long pink legs make identification simple. This bird breeds in the south-east and south-west, central France and Hungary, and recently in Holland also.

An odd-looking wader is the Pratincole (*Glareola pratincola*), a summer visitor to parts of the south. It looks remarkably like a large swallow in flight and feeds on insects which it often catches in the air over very hard and dry areas of mud or lake margins.

Skuas (family Stercorariidae) are pirates, chasing other birds, especially gulls, until they drop or disgorge the fish which they have caught. The Arctic Skua (*Stercorarius parasiticus*) breeds

in colonies along coasts of the far north and Arctic. There are three colour phases, light, intermediate and dark, and all have the central tail feathers elongated so that they extend about three inches beyond the rest of the tail.

Gulls and terns (family Laridae) are represented in Europe by many species. The largest is the Great Black-backed Gull (*Larus marinus*), breeding in Iceland and along the northern coasts. In winter it is also seen as far south as the coasts of Spain, and during the summer it frequently steals chicks and eggs from the nests of other gulls and seabirds. The Glaucous Gull (*Larus hyperboreus*) is about the same size as the Great Black-back and is white with a pale grey back.

A familiar gull is the Black-headed Gull (*Larus ridibundus*) whose chocolate-brown head changes in autumn to white with a dark patch behind the eye. It breeds over most of Europe from France eastwards, nesting in large colonies on coastal marshes and moors, and is often seen following ploughs, picking up worms.

The Kittiwake (*Rissa tridactyla*) is a sea and coastal bird, breeding in large numbers on steep cliffs around the coasts of Iceland, the British Isles and Brittany and Norway. This is a delicately coloured gull with an elegant and buoyant flight. At breeding colonies the call 'kitti-wek' is heard regularly.

Whereas gulls are generally rather heavy-looking birds, terns are more graceful and slim. All are summer visitors, and are distinguished from gulls by the deeply forked tail in most species. The Common Tern (*Sterna hirundo*) breeds in large colonies on marshes, islands and dunes. It sports

Above far left: wherever thistles abound Goldfinches (Carduelis carduelis) *are likely to be seen.*
Above centre left: the Nightingale (Luscinia megarhynchos) *builds its nest in undergrowth near the ground.*
Above left: the lively Blue Tit (Parus caeruleus).
Left: in Britain the Robin (Erithacus rubecula) *is very confiding to man.*
Right: Tengmalm's Owl (Aegolius funereus).

the typical tern plumage of white, with pale grey back and wings, and black cap extending to the nape of the neck. The bill is red with a black tip. The very similar Arctic Tern (*Sterna paradisea*) breeds in the north and travels to Antarctic waters to winter. The beautiful Roseate Tern (*Sterna dougallii*) with its exceptionally long tail streamers breeds locally among colonies of Common and Arctic Terns in the north.

The Caspian Tern (*Hydroprogne tschegrava*) at twenty-one inches is the largest and most aggressive of the terns. On breeding grounds terns will dive at any unwelcome visitors, including humans, and their long, sharp bills are capable of drawing blood.

While these terns are all mostly marine birds, the members of the genus *Chlidonias* prefer marshes and lakes and indeed are often referred to as marsh terns. The Black Tern (*Chlidonias niger*) builds its floating nest in colonies in southern Spain and Portugal and from France eastwards. Unlike other terns it prefers insects as food, picking them off the water surface.

Auks (family Alcidae) are black and white, penguin-like birds, breeding on cliffs in large colonies along the northern coasts. The commonest member of the family also nests on the north coast of the Iberian peninsula. This is the Guillemot (*Uria aalge*). Some birds have a narrow white ring around the eye and a white line extending backwards from this. These are called 'bridled' forms and are fairly common in some places. Huge colonies consist-

Right Barn Owl (Tyto alba) *returning to its nest with a vole.*
Below: the huge Eagle Owl (Bubo bubo).

ing of this species and the Razorbill (*Alca torda*), often joined by Kittiwakes and Puffins (*Fratercula arctica*), are a wonderful sight in June when scarcely any space on the cliff face is unoccupied.

The unmistakable Puffin with its huge red, blue and yellow bill, orange feet and large head looks quite a comical bird. It nests in a small burrow or in a natural hole or crevice, and has a habit of standing at the entrance, surveying the scene like a wise old man. Its bill is capable of carrying half a dozen or more fish. In late summer auks desert the coasts and spend the winter at sea.

Of the family Pteroclidae, the sandgrouse, none of the species is likely to be seen by any but the most dedicated bird watcher. The Black-bellied Sandgrouse (*Pterocles orientalis*) breeds in Iberia and Cyprus; the Pin-tailed (*Pterocles alchata*) in the south of France and

north and west and from being a nine days' wonder in England in 1952 is now almost as much of a pest there as the Woodpigeon (*Columba palumbus*).

The Cuckoo (*Cuculus canorus*) is a widespread breeding species known to everyone by its call. The Great Spotted Cuckoo (*Clamator glandarius*) is a common passage migrant and summer visitor in Cyprus where it also breeds.

Of the thirteen species of owl, three—the Snowy Owl (*Nyctea scandiaca*), Hawk Owl (*Surnia ulula*) and Great Grey Owl (*Strix nebulosa*)—are confined to the extreme north of the continent, whereas the small Scops Owl (*Otus scops*) breeds mostly in the southern half except in Russia where it reaches the Leningrad district. Its monotonous 'pew' note often gives away its presence after dusk on buildings in towns in the south. The Eagle

Above: newly hatched Cuckoo (Cuculus canorus) *ejecting egg from Tree Pipit's nest.*
Top: in spring male Ruffs (Philomachus pugnax) *sport huge ruffs and ear-tufts.*
Right: young Cuckoo in nest of Reed Warbler.

Iberia. Sandgrouse inhabit dry semi-deserts and nest on the ground.

On the other hand, the family Columbidae (pigeons and doves) are widespread in Europe, although numbers in Russia have declined enormously owing to the forests being felled and the birds being shot for food. The most interesting is the Collared Dove (*Streptopelia decaocto*) which in the last eighteen years has extended its breeding range from south-east Europe to the

are the Bee-eater (*Merops apiaster*), a regular breeder in southern Europe and occasionally further north, the Roller (*Coracias garrulus*) whose main breeding area is to the east and south-east from France, and the Hoopoe (*Upupa epops*) which is widespread over the mainland of Europe except Scandinavia. All these three species breed in holes.

Europe has no fewer than ten woodpeckers including the Wryneck (*Jynx torquilla*), so-called from the extraordinary contortions of its neck when it is alarmed. Its plumage is mottled

Left: unlike the female, the male Nightjar (Caprimulgus europaeus) *has white patches on its wings and tail.*
Above: a Nightjar can be seen yawning, showing its enormous gape.
Right: Sand Martin (Riparia riparia) *feeding its young at the entrance to its nest burrow in a sandy cliff.*

Owl (*Bubo bubo*) is the largest European species, a truly magnificent bird with large orange eyes and prominent ear-tufts. Unfortunately it is now scarce in many parts of its range through persecution.

The nightjars are a curious family, entirely insectivorous. The Common Nightjar (*Caprimulgus europaeus*) is found over the whole continent, whereas the Red-necked Nightjar (*Caprimulgus ruficollis*) is confined to the Iberian Peninsula as a breeding species; both birds are beautifully marked in shades of brown and grey which afford excellent camouflage as they sit on the ground during the day.

Swifts (family Apodidae) are represented by three species of which *Apus apus* will be very familiar to most people as the flocks fly very fast, screaming among roof-tops in summer, dark brown birds with scythe-shaped wings larger than Swallows (*Hirundo rustica*). The other two species, the Pallid Swift (*Apus pallidus*) and the Alpine Swift (*Apus melba*), are confined to the south. The familiar Swift nests in roofs of houses in the British Isles but in Russia it often uses holes in trees in forests.

The brilliantly coloured Kingfisher (*Alcedo atthis*) only too often appeared stuffed in glass cases in unenlightened Victorian days. It was very badly hit all over its range in the severe winter of 1962–3 but is making a slow recovery. The Pied Kingfisher (*Caryle rudis*) has occurred in Greece, Poland and Cyprus, and the Belted Kingfisher (*Caryle alcyon*) from North America has been recorded from Holland and Iceland. Three more brilliantly coloured birds

like a nightjar's, and it is a summer visitor to most of Europe but is more often heard than seen. The most striking of the actual woodpeckers is the Black Woodpecker (*Dryocopus martius*), a bird as large as a Rook (*Corvus frugilegus*) and all black except for crimson on the head. The Greater-spotted (*Dendrocopus major*), Middle-spotted (*Dendrocopus medius*) and Lesser-spotted Woodpeckers (*Dendrocopos minor*) as also the Syrian Woodpecker (*Dendrocopos syriacus*) and the White-backed Woodpecker (*Dendrocopos leucotos*), are all a mixture of black, white and red and are not easily distinguishable one from the other in the field. They are all smaller than the Green Woodpecker (*Picus viridis*) and the Grey-headed Woodpecker (*Picus canus*) which somewhat resemble one another.

All the rest of the European birds belong to the large order Passeriformes. Of the family Alaudidae (larks) the Calandra Lark (*Melanocorypha calandra*) and the Short-toed Lark (*Calandrella brachydactyla*) breed in the extreme south, but the Skylark (*Alauda arvensis*) and Wood Lark (*Lullula arborea*) are commonly present over most of Europe except the extreme north. Some are famous for their outstandingly beautiful song, especially the Skylark (chosen incidentally by the Danes as their 'national bird') and the Wood Lark. Except for the conspicuous male Black Lark (*Melanocorypha yeltoniensis*) which is rarely seen west of Russia, larks' plumage is mostly of rather dull shades of mottled brown.

The Sand Martin (*Riparia riparia*), Swallow (*Hirundo rustica*)—chosen as their 'national bird' by the Estonians—and House Martin (*Delichon urbica*) breed in the whole area except Iceland. The last two are familiar to country people through their habit of nesting in farm buildings and on houses and through their cheerful twittering.

There are seven European pipits. Of these the Red-throated Pipit (*Anthus cervinus*) breeds only in the far north of Norway and Lapland. The other pipits are by no means always easy to distinguish one from another; often the voice is the best guide, especially in the

case of the Tree Pipit (*Anthus trivialis*) which ends its song with a parachute descent to a perch, calling a characteristic 'seea-seea-seea'. The Yellow Wagtails (*Motacilla flava*) and their various races are widespread in the continent, each race in its own breeding area. They are dainty birds and the male of the British race (*Motacilla flava flavissima*) is particularly beautiful and so is that of the Black-headed Wagtail (*Motacilla flava feldegg*) which breeds in Cyprus.

The Grey Wagtail (*Motacilla cinerea*) is larger than other wagtails, and the male in summer is a beautiful mixture of grey-blue, black, yellow and white. The Pied and White Wagtails (*Motacilla alba*) are common and ubiquitous. Very large roosts occur in autumn and winter in reed beds and greenhouses. These are cheerful birds which run very fast but do not hop.

There are five species of shrikes, often known as butcher birds from their habit of impaling their prey (such as small birds, lizards, and beetles) on thornbush 'larders'. As a breeding species the Masked Shrike (*Lanius nubicus*) is confined to the south-east. The Red-backed Shrike (*Lanius collurio*) has been decreasing in Britain and north-west Europe for at least twenty years, possibly because of drainage of its wild habitats.

The very handsome and striking Waxwing (*Bombycilla garrulus*), so called from the scarlet waxy tips to the secondaries, is confined as a breeding species to the far north of Finland, Scandinavia and Russia but irrupts periodically in winter into central and north-west Europe. As it is usually indifferent to human beings one may often admire it at close quarters even in Britain in years of irruption.

The Dipper (*Cinclus cinclus*) may be seen bobbing on a rock in swift hill

Top: the caressing display of the Wood Pigeon (Columba palumbus); *the male is on the right.*
Left: Great Grey Shrike (Lanius excubitor) *at its 'larder' of impaled victims.*
Right: the Black Woodpecker (Dryocopus martius).

streams or even walking under water; its large nest is often built in a crevice behind a waterfall. Its breeding range extends over most of Europe except Iceland, but in central European Russia its status is obscure. A slightly different race which lacks the chestnut edge to the white breast has been chosen as Norway's 'national bird'.

Europe has only one Wren (*Troglodytes troglodytes*), but that breeds and winters everywhere except in the far north of Scandinavia. It is the continent's smallest bird apart from the Goldcrest (*Regulus regulus*) and Firecrest (*Regulus ignicapillus*) and has a loud song for its size. It sings practically all the year round; almost any habitat seems to suit it.

Of the three accentors (family Prunellidae) the Alpine Accentor (*Prunella collaris*) breeds only on rocky mountain slopes up to the snow-line; the Dunnock (*Prunella modularis*) is common and widespread except in Iceland and the far south, where, however, it is a winter visitor from central Europe; and the Siberian Accentor (*Prunella montanella*) is accidental in Czechoslovakia, Greece, and Italy. The eggs of the Dunnock are a very lovely blue.

The Muscicapidae (warblers, flycatchers, thrushes, etc.) are a very large family; warblers alone number thirty-nine species in Europe. Many are by no means easy to separate by differences of plumage visible in the field. They may be divided into swamp warblers, scrub, leaf and tree warblers. In nearly every case the song is a great help to identification. Swamp warblers are mostly brown, some with a prominent eye stripe; leaf warblers are mostly some shade of green or yellow; scrub warblers are more distinctively coloured, many of them with a capped appearance. The Reed Warbler (*Acrocephalus scirpaceus*) is the most widely spread of the swamp warblers and is a colonial breeder.

Of the scrub warblers the Garden Warbler (*Sylvia borin*), Blackcap (*Sylvia atricapilla*) and Whitethroat (*Sylvia communis*) breed over most of Europe except in the north of Scandinavia and, in the case of the Garden Warbler, in those parts bordering on the Mediterranean; the first two are noted songsters. The skulking Dartford Warbler (*Sylvia undata*) is the only one breeding in England that does not migrate and consequently it was almost exterminated there in the severe winter of 1962–3. It probably also has to contend with egg-collectors, whose activities, it need hardly be said, are quite illegal.

One of the scrub warblers worthy of mention is the Subalpine (*Sylvia cantillans*) which is a colourful summer visitor to most of Spain, Italy, the Dalmatian Coast, Greece and the Mediterranean islands. Another is Ruppell's Warbler (*Sylvia rueppelli*) of which the male has a striking black crown, face and throat set off by a conspicuous white moustachial stripe. It breeds only in the Aegean region.

The three commonest leaf warblers are the Willow Warbler (*Phylloscopus trochilis*), the Chiffchaff (*Phylloscopus collybita*) and the Wood Warbler (*Phylloscopus sibilatrix*). One or other occurs over most of Europe except that they are all poorly represented in the Iberian Peninsula. The first two look almost identical but their songs are utterly different, the former's being a sweet plaintive cadence and the latter's a monotonous 'chiff chaff', whence its name. The Wood Warbler, though not dissimilar, is a rather larger and more strongly coloured bird with a remarkable song 'piu piu piu' followed by a loud shivering trill. When heard among the new green beech or oak leaves in early spring and even more when also seen high up in the foliage this bird

offers an annually recurring delight to the watcher. All three species build well-concealed nests on or very close to the ground.

The Goldcrest and Firecrest look alike and are the smallest European birds. The Goldcrest is the 'national bird' of Luxembourg, being a symbol of that small country. Neither species occurs in Iceland nor along the east coast of Spain but most of the rest of Europe holds one or the other. Both are mainly woodland birds and build nests suspended from branches. The Goldcrest has a yellow line down the centre of the crown; the Firecrest's is golden. Their songs are so thin and high-pitched that they are inaudible to some people.

None of the four flycatchers in Europe occurs in Iceland. The Spotted Flycatcher (*Muscicapa striata*) breeds in the whole of the rest of the continent as far east as the Urals. The next most widely distributed is the Pied Flycatcher (*Ficedula hypoleuca*) but it is absent from Ireland, the east side of Great Britain, much of France and Portugal and from Italy and south-east Europe where its place is largely taken by the Collared Flycatcher (*Ficedula albicollis*). The Red-breasted Flycatcher (*Ficedula parva*) only breeds roughly on a level with France to the east. Apart from the orange throat of the male of this species the others are all brown and white or black and white.

The Whinchat (*Saxicola rubetra*) and Stonechat (*Saxicola torquata*) are rather similar birds, the former always distinguishable by a white eye stripe and some white in the tail. Both are birds

Left: Starlings (Sturnus vulgaris) *settling down to roost at dusk.*
Above: Hawfinch (Coccothraustes coccothraustes) *bathing. Its massive bill is used to crack open nuts and large seeds.*

of open country with bracken and bushes of gorse, and both nest on the ground.

There are seven wheatears in Europe. By far the most ubiquitous is the Common Wheatear (*Oenanthe oenanthe*) which breeds in suitable habitats over the entire continent except Cyprus. All are about six inches long and have contrasting black, white and, except for the handsome Black Wheatear (*Oenanthe leucura*), buff plumage. All inhabit open, usually barren, ground and breed in holes in rocks or walls. The Black-eared Wheatear (*Oenanthe hispanica*) which breeds throughout

southern Europe is an extremely handsome bird.

The Rock Thrush (*Monticola saxatilis*) and the Blue Rock Thrush (*Monticola solitarius*) are beautiful birds, about eight inches long, slightly smaller than a Song Thrush (*Turdus philomelos*). Both are confined to rocky regions in the south and utter their fluty warbles in a vertical display flight.

The females are speckled brown like female Blackbirds (*Turdus merula*) but the male Rock Thrush has a blue head and shoulders and brilliant orange breast and tail, whereas the Blue Rock Thrush is a rich dark blue with black wings and tail. The Blackbird is Sweden's 'national bird' as its late winter song is considered to herald the arrival of spring.

Europe has two redstarts. Neither breeds in Iceland, Ireland or Cyprus, but the Redstart (*Phoenicurus phoenicurus*) breeds in most of the rest of Europe except Dalmatia and Greece, and the Black Redstart (*Phoenicurus ochruros*) breeds almost everywhere else except in Scandinavia, Great Britain (though occasionally in the south of England), Finland and European Russia except the south-west. Both nest in holes, and in central Europe the Black Redstart tends to take the place of the Robin (*Erithacus rubecula*). The Robin, always a favourite with the British, partly because of its confiding attitude towards man in the British Isles, has been voted Britain's 'national bird'. It breeds everywhere except in the northern half of Scandinavia, where it is replaced by the Bluethroat (*Cyanosylvia svecica*), the east coast of Spain, and in Cyprus where it is an abundant winter visitor.

The Nightingale (*Luscinia megarhynchos*) is surely one of the most famous of European birds. It is nothing much to look at but its song has been admired down the ages and the effect is enhanced by its being sung by night as

Top left: the Long-tailed Tit (Aegithalos caudatus) *builds a beautiful nest of feathers.*
Left: the Lesser Whitethroat (Sylvia curruca) *is usually a very shy bird.*

well as by day. It is widespread over Europe except Iceland, Ireland, Great Britain north of a line joining the Severn and the Wash, Scandinavia and eastwards from north-east Germany where it is replaced by the Thrush Nightingale (*Luscinia luscinia*) whose song is similar but, in the opinion of some, slightly inferior.

The Red-flanked Bluetail (*Tarsiger cyanurus*) has been spreading westwards since the late 1930s as a breeding species in the far north of European Russia and has reached eastern Finland; elsewhere it is a rare straggler. It inhabits dense pine or spruce forests. It is a strikingly coloured bird about the size of a Redstart; the juvenile resembles a young Robin.

Seventeen species of *Turdus* occur of which no fewer than eleven are vagrants. The distribution of the Ring Ouzel (*Turdus torquatus*) is patchy on hilly moorlands. The other five are all common in winter as partial migrants from the north and east of the area, even as far as Iceland in the case of the Fieldfare (*Turdus pilaris*) and Redwing (*Turdus iliacus*), and from their breeding areas which cover most of the continent in the cases of the Blackbird, Song Thrush and Mistle Thrush (*Turdus viscivorus*). These last three all have very pleasant songs, the Mistle Thrush delighting to sing in stormy weather from a high perch. The song of the Song Thrush has been brilliantly described in Browning's well-known lines:

'*That's the wise thrush; he sings
    each song twice over,
Lest you should think he never
    could recapture
The first fine careless rapture!*'

There are eleven species of tits, none of which occurs in Iceland. These are all small restless birds, acrobatic in behaviour, as happy hanging upside-down as when perched more normally. The Penduline Tit (*Remiz pendulinus*) and the Long-tailed Tit (*Aegithalos caudatus*) build most exquisite suspended ovoid nests from hundreds of feathers, with a tiny entrance at the top; the others build in natural holes or nestboxes. Particularly well known and widespread species are the Marsh Tit (*Parus palustris*), Coal Tit (*Parus ater*), Blue Tit (*Parus caeruleus*) and Great Tit (*Parus major*).

The Nuthatch (*Sitta europaea*) is generally distributed except in Iceland, Ireland, Scotland and to the north of southern Scandinavia and in Finland. The Corsican Nuthatch (*Sitta whiteheadi*) is confined to Corsica and very rare even there, and the Rock Nuthatch (*Sitta neumayer*) is confined to Dalmatia and Greece. These birds nest in holes and plaster the entrance with mud which sets very hard. They differ from woodpeckers in their ability to move on trees downwards as well as upwards and without using the tail as a support.

The Wall Creeper (*Tichodroma muraria*) is a mountain bird, usually above

Right: Tree Creeper (Certhia familiaris).
Below: Bearded Reedling (Panurus biarmicus).

Above: male Redstart (Phoenicurus phoenicurus).
Centre: Whinchat (Saxicola rubetra).
Far right: Fieldfare (Turdus pilaris).

6,000 feet, confined to the Pyrenees, Switzerland, the Balkans, Carpathians and Cyprus where it has been seen some fifteen times in winter; it is a strikingly coloured bird with grey head and back and bright red wings, usually seen fluttering over a rock face.

The Tree Creeper (*Certhia familiaris*) is resident in the British Isles, south Scandinavia and eastern and south-eastern Europe, and the Short-toed Tree Creeper (*Certhia brachydactyla*) in western Europe and Iberia but not north of the English Channel nor in Sardinia, Sicily or the Peloponnese. They spend their lives hunting insects in the crevices of boles of trees, beginning from the bottom and working upwards. Their fragile nest is usually made behind bits of loose bark.

Buntings are represented in Europe by fifteen species one of which, the Pine Bunting (*Emberiza leucocephala*), does not breed west of east European Russia. Of the remaining fourteen, only the Snow Bunting (*Plectrophenax nivalis*) breeds in Iceland as well as in Norway. The Yellowhammer (*Emberiza citrinella*), famous for its song 'A little-bit-of-bread-and-no-CHEESE', is stated to be in enormous numbers in Russia, and the Red-headed Bunting (*Emberiza bruniceps*) is one of the most common birds in the southern part of the area between the Volga and the Urals. The most handsome of all the European buntings is the male Black-headed Bunting (*Emberiza melanocephala*) but, alas, it breeds only along the shores of the Adriatic, in Greece and its islands and in Cyprus; its pleasant song is often delivered from the very top of a tree.

There are twenty-one finches in Europe of which two are only vagrants. The Redpoll (*Acanthis flammea*) is the only one to occur in Iceland where it breeds and winters. The very widely distributed ones are the Chaffinch (*Fringilla coelebs*), Greenfinch (*Carduelis chloris*), Goldfinch (*Carduelis carduelis*), Bullfinch (*Pyrrhula pyrrhula*) and Hawfinch (*Coccothraustes coccothraustes*). This last is a top-heavy looking, large finch with a very powerful bill able to crack the hardest fruit stones to get at the kernels; it is a very shy species, rarely seen even where it is present. The Goldfinch and the Linnet (*Acanthis cannabina*) are unfortunately favourite cage-birds in several countries.

The House Sparrow (*Passer domesticus*) and Tree Sparrow (*Passer montanus*) are both very widely distributed over the continent, the former being familiar to everyone as it seldom lives far from human habitation; the latter differs from it, apart from the plumage, by the fact that the sexes are alike whereas they are quite different in the House Sparrow.

The three starlings are the Rose-coloured Starling (*Sturnus roseus*), a strikingly pink and black bird when adult, which breeds irregularly as far west as Italy and is highly esteemed in Russia because of its habit of feeding on locusts; the Common Starling (*Sturnus vulgaris*) which breeds regularly almost everywhere except in Iberia, the

Above: the Reed Warbler (Acrocephalus scirpaceus) *builds an intricate suspended nest in reed-beds.*
Centre: male Pied Flycatcher (Ficedula hypoleuca), *its bill full of grubs.*
Far right: the slender Tree Sparrow (Passer montanus).

Mediterranean islands and Greece, and congregates in vast flocks at roost in autumn and winter on city buildings and in woods and reed-beds; and the Spotless Starling (*Sturnus unicolor*). Starlings are cheerful, ungainly birds with a varied song delivered from the tops of trees and roofs.

The family Oriolidae has only one representative in Europe, the Golden Oriole (*Oriolus oriolus*). As the cock bird is a glorious contrast of yellow and black and sings a lovely fluty whistle, it would be better known if it were not so much given to keeping to thick foliage where it is difficult to pick out among sunshine and shade.

The crows (*Corvidae*) are the largest and perhaps from a human viewpoint the most 'intelligent' of the perching birds. The Jay (*Garrulus glandarius*), Magpie (*Pica pica*) and Jackdaw (*Corvus monedula*) are the most widely distributed as breeders though none of them occurs in Iceland—nor the Jay and Jackdaw in the north of Scandinavia, being replaced there by the Siberian Jay (*Perisoreus infaustus*), the Hooded Crow (*Corvus corone corvix*) and the Raven (*Corvus corax*). The Rook (*Corvus frugilegus*), Carrion Crow (*Corvus corone*) and Raven are all glossy black and the Jackdaw is also black but with a grey nape and pale grey eyes. Hooded Crows, a subspecies of the Carrion Crow with which they interbreed, migrate in October in vast numbers accompanied by large numbers of Rooks and Jackdaws, all following the coast line along the south side of the Gulf of Finland and the Baltic and over the Kurische Nehrung.

An interesting corvine is the Nutcracker (*Nucifraga caryocatactes*) of which an unprecedented invasion into Great Britain occurred from August to October 1968, in far greater numbers than the total of all previous years; all the countries of north Europe from Russia, Poland, the Baltic States and Scandinavia to Belgium and northern France were concerned in this mass movement. The bird is about a foot long, chocolate brown speckled with white, and with broad black wings. It breeds in the mountains of Switzerland, Dalmatia and the Carpathians as well as in the north-east of the continent.

The Raven is, so to speak, top of our current taxonomic tree, which starts at the bottom with the species considered the most primitive and works upwards. Dr Konrad Lorenz tells fascinating stories of his tame Ravens in his book *King Solomon's Ring*, and Gilbert White in *The Natural History of Selborne* wrote:

'They spend all their leisure time in striking and cuffing each other on the wing in a kind of playful skirmish; and, when they move from one place to another, frequently turn on their backs with a loud croak, and seem to be falling to the ground. When this odd gesture betides them, they are scratching themselves with one foot, and thus lose the centre of gravity.'

MAMMALS
Cathy Jarman

Of the nineteen major living orders of mammals in the world only eight are found in Europe and these include the seals and whales found around its shores.

The mammals inhabiting the temperate forests of Europe seem to be far less varied and numerous than those of the tropical countries, mainly as the result of man's increased population and its effect on the European environment. In western Europe in the last 1,000 years there has been a large population increase. With this growth, the natural vegetation changed as towns grew and industry and agriculture became more intensified. This led to the destruction of thousands of acres of natural forests. The only little-industrialized areas of Europe today are to be found in the Balkans, Russia and Siberia, and here the condition of the forests remains fairly unchanged.

There are comparatively few large mammals surviving in Europe, most of the larger species having been exterminated in the Middle Ages. In the last century also there has been a tremendous acceleration in animals becoming extinct. The European Bison or Wisent (*Bison bonasus*) would be extinct now if naturalists and conservationalists had not rescued the remaining herd after the Second World War. As it is, the Wisent survives only in protected herds in captivity. In 1914 nearly 1,000 animals lived in the forests of Bialowieza on the Polish–Russian border, but by the end of the First World War, all these had been killed as a result of bombing and slaughtering for food. Another group in the Caucasus were also all dead by 1927. Later, a small number of bison were purchased from other sources and a herd gradually re-established. In 1956 several bison were released from confinement and now fifty-seven live in the Bialowieza forest in a wild state, of which thirty-four were born in the wild. The result of this acclimatization of bison to wild

The European Rabbit (Oryctolagus cuniculus) *is very agile but not as fast as the hare which has longer legs. Its white tail bobs when it runs.*

conditions has been very satisfactory. The animals have retained good condition and breeding is carried on normally. The Polish authorities are very anxious to distribute the bison as widely as possible in suitable areas, but the only areas suitable with enough forested parts are in the Soviet Union.

Because it has been extinct in the wild for many years, little is known of the Wisent's behaviour. Compared to the American Bison (*Bison bison*) it has longer legs, is longer in the body and has a smaller head which it carries much higher. The single, yellow-coloured young are born in May or June after a gestation period of 260-270 days. The bulls are mature between six and eight years, the cows earlier at three or four years of age.

Other hoofed mammals ranging the temperate forests are the Red Deer (*Cervus elaphus*), the Roe Deer (*Capreolus capreolus*), and the Fallow Deer (*Dama dama*). Originally the Red Deer lived in very open woodland but in certain areas such as Scotland they have become completely adapted to living on the higher moorlands. The herds are on the whole nocturnal, usually resting during the day. Few obstacles prevent their movements, as they are expert jumpers and can swim any river or lake with ease. Although they live in herds, the sexes are separated for much of the year. The composition of the herds varies with the time of the year. Outside the breeding season or rut, the hinds and young live in herds under the leadership of an old hind. The stags are rarely found with these herds. During the rutting season, however, the males collect a 'harem', using their mating call of a long-drawn-out loud and deep bellowing and roaring. The hind makes a sharp bark when anxious and also a growling sound to her young. The male outside the rut is fairly silent, only barking like the female when threatened by an intruder.

Roe Deer are more numerous, but prefer young woods or those with dense undergrowth. They also occur in very moist areas and do tend to go high into the mountains. They are smaller than the Red Deer and have smaller antlers.

The attractive Fallow Deer, with its bright fawn coat and the large white spots on its back and sides, originally came from the Mediterranean forests but is now very common in the parks and forests of Europe. The male has a prominent 'Adam's apple' and the rutting cry is a deep-toned grunt. The young, usually one, occasionally two, are born in June and July, and they are only slightly spotted. Antlers appear in the young males in their second year, at which stage they are known as 'prickets'.

The only other 'big game' animal to survive in Europe is the Wild Boar (*Sus scrofa*). This burly, dark-coloured and bristly animal with large upward-curving tusks, prefers to live in the deciduous forests near to small lakes, marshes and arable land. It is mainly nocturnal, but loves to stretch out in the

Left: the Musk Ox (Ovibos moschatus), *which occurs in Greenland and north-east Canada, has been reintroduced into northern Norway.*
Right: the European Bison or Wisent (Bison bonasus).

sun during the day. Its penetrating smell is somewhat offensive, and its snorting is a familiar noise. The Wild Boar has large families; there may be as many as twelve fawn and white striped piglets in a litter, born after a gestation period of about 115 days.

Compared with the carnivores, the hoofed mammals have adapted extremely well to the modern agricultural situation. Brown Bears (*Ursus arctos*) used to be widespread throughout Europe, but they survive now only in a few little-frequented districts; in western Europe they are found in small pockets in the Pyrenees and in Scandinavia, and in larger numbers in eastern Europe.

These ambling mammals often cover long distances and on their travels love to bathe and swim. Although classed as flesh-eaters in the order Carnivora, Brown Bears are omnivorous, eating all kinds of vegetable matter, honey from wild bees' nests, carrion, fish or animals they kill for themselves. When hunting they are fairly agile, and kill not by hugging but by slashing powerful blows with their forearms. Their paws are armed with a strong set of long unretractile claws.

The European Wild Cat (*Felis silvestris*), looking like a large domestic tabby, inhabits the most remote wooded areas in Scotland, France, Central Europe, Italy, Greece and a large part of Russia. The den is usually inside a hollow tree, burrow or among rocks. Here up to six kittens are born in the spring after sixty-eight days gestation. Immensely playful and attractive like all kittens, they soon grow and by autumn they are almost fully grown and able to fend for themselves. Hunting methods are typical of the cat family; after a long patient wait the

Top left: the antlers of the male Elk or Moose (Alces alces) *may reach a width of over 6 feet and are used as a protection against predators.*
Left: the Red Deer (Cervus elaphus) *inhabits temperate forests of Europe.*
Right: during the rutting season male Red Deer (Cervus elaphus) *will engage in a show of strength to try to establish dominance.*

prey is caught ambush style or, after stealthy stalking, a victim is brought down by a quick, silent pounce.

The Red Fox (*Vulpes vulpes*) hunts at night for small rodents or birds, keeping a sharp lookout for any danger. As it tracks down a scent trail it keeps its nose to the ground, but at other times its head is held erect, with the thick, long, bushy tail held downwards, as it trots in a characteristic manner. The Red Fox is usually thought of as doing great damage to farmers by stealing their hens and causing chaos in the farmyard. It has, however, several points in its favour; it eats mice and shrews, even digging to get these tasty meals; insects such as grasshoppers, dung-beetles and unwanted pests are taken; and it will also eat any berries or fruits which fall to the ground. The fox digs its underground home or den in a suitable hidden spot, in woods, undergrowth or thickets, often enlarging a rabbit's hole or taking over a disused badger set, although it has been known to live amicably with the badger in the same set. There are usually several entrances to the fox's den: as a rule there is a regular arrangement with a cavity near the entrance, further down a storage chamber and finally, deep down, a nest chamber. Here the vixen gives birth to five to eight young in April and both parents share in bringing food back to the rapidly growing young. The cubs are blind at birth, their eyes opening at ten days. After four weeks they venture to the outside world, where they can be seen having playful fights and investigation trips. By August they are ready to depart and lead independent lives.

The small mammals have suffered much less than the larger ones, and in some cases the changed vegetation has helped to increase their numbers. Of the insectivores, the European or Spiny Hedgehog (*Erinaceus europaeus*), the Algerian or Vagrant Hedgehog (*Erinaceus algirus*), and the Common Eurasian Mole (*Talpa europaea*) are fairly abundant. The hedgehog predominates in dry areas, in shrubs, hedges, copses and fringes of woods, having settled in the changed environment. The thickset mole is found in a wide range of habitats in almost any kind of soils, from open fields almost to the sea coast and from woods to the mountain lowlands. The only soil it avoids is that of pure sand. By day the hedgehog rests in the undergrowth or thick hedgerows. Its sleeping place may be detected at times by the loud snores it makes. At night it patrols its territories searching for insects, slugs, snails, worms, and other small animals. It is true that it will kill snakes, but only occasionally. The hedgehog's main predators are men and dogs, but there is definite evidence that foxes and badgers do not have

Above: the Red Fox (Vulpes vulpes) *is mainly nocturnal but may be seen during the day in places where it is undisturbed.*

Below: a one-week-old Roe Deer (Capreolus capreolus).
Right: male Fallow Deer (Dama dama) *grazing.*

much trouble in opening up hedgehog, and often leave the skin turned inside out.

The Mediterranean or Blind Mole (*Talpa caeca*), so called because often its eyes are hidden beneath the skin, occurs throughout the Mediterranean countries, while the Roman Mole (*Talpa romana*) is found in southern Italy, Sicily, Corfu and Greece.

Distributed throughout Europe are eleven species of shrews. Of these small mouse-shaped mammals with long pointed snouts, the Common Shrew (*Sorex araneus*), the Water Shrew (*Neomys fodiens*) and the Pygmy Shrew (*Sorex minutus*) are the most widely distributed. The Pygmy Shrew is the only shrew found in Ireland.

The Common Shrew measures only three inches long and weighs usually only about a third of an ounce. Although only small they are extremely fierce and will attack and eat large insects, such as beetles and worms, with relish. They will also attack and devour any other small animal they meet. The Common Shrew is a swift and bustling mammal, exploring busily with twitching snout and whiskers, and every now and then it will rear up and appear to sniff the air.

Except during the breeding season and when the young are still with the mother, shrews are solitary and aggressive. If a neighbouring shrew or newcomer approaches, a vicious fight will ensue; first they scream at each other, then perform threatening postures and last of all resort to biting, but they rarely wound one another severely.

Common Shrews only live for about fifteen months. The female at a year old bears her litter of about five in a nest of dried leaves. The adult population, however, do not live to the following year but die in late summer or early autumn following the breeding period.

Rodents are numerous throughout the world and Europe is well endowed with them, having over fifty species. Rodents are characterized by the possession of two large long upper incisors. Squirrels, dormice, hamsters, mice and rats are all included in this order which is the most abundant order of all mammals. Most of them live secretive and nocturnal lives.

One of the more attractive of the rodents is the Red Squirrel (*Sciurus vulgaris*) which is found throughout wooded parts of Europe from the tree line south to the Mediterranean coast. Numbers have been reduced in parts of Britain by an epidemic, and also in certain areas it has almost completely disappeared due to the introduction of the American Grey Squirrel (*Sciurus carolinensis*). The Red Squirrel does not hibernate throughout the winter as many believe; during this cold season it is active very early in the day. It moves with agility and speed when travelling through the trees, jumping great distances from branch to branch, and descending tree trunks head first. As these squirrels go through their daily routine, various calls are made, from chattering to growling and wailing. During a chase fierce scolding and whistling sounds occur.

The American Grey Squirrel is larger and more robust than the Red Squirrel, and does not have ear-tufts. In summer it is usually brownish-grey with bright, rufous hair on paws and flanks and a white belly. In winter it is greyer with a brownish stripe running down the back, and the tail hairs are fringed with white. It inhabits mixed deciduous woodland and favours oak and beech trees. It was introduced into Britain between 1876 and 1920 and has since

spread from various centres to most parts of Britain, including Ireland. It is still rare if not entirely absent in the Highlands, Cumberland, Norfolk and the Isle of Wight, which are still strongholds of the Red Squirrel; and it is still absent from continental Europe.

The mice and rats are extremely widespread and the mice are most difficult to identify. The House Mouse (*Mus musculus*) is quite small and stockily built with a sharply pointed muzzle, whereas the Harvest Mouse (*Micromys minutus*) has a smaller-looking head and ears and a blunter snout. The size of the House Mouse varies but it is usually around three inches in length with a slightly longer tail, and it weighs about an ounce. Its

Top: the Reindeer (Rangifer tarandus) *is unique among deer in that the male and female are both antlered.*
Left: the antlers of the Fallow Deer (Dama dama) *are broadly palmated.*
Above: young Roe Deer (Capreolus capreolus).

*Left: a large cat which was relatively common until recent times is the Northern Lynx (*Felis lynx*).
Above: although extinct in Britain, the Wolf (*Canis lupus*) is still common in eastern Europe, Italy and Spain.
Right: the European Wild Cat (*Felis sylvestris*) *prefers deep forests in uninhabited regions.*

name refers to its close association with man, as it lives in dwellings, food stores and farm buildings such as stables and barns. In summer it may spread to hedgerows and gardens where it competes with the Wood Mouse (*Apodemus sylvaticus*).

The success of the House Mouse is mainly due to its adaptability; it breeds almost anywhere provided it is undisturbed and food and shelter are at hand. It has even been found living in refrigerated stores and down mines. Large populations can be quickly established in favourable areas, as between five and ten litters can be produced by one female in a year. The breeding season lasts from early spring to late autumn, between four and eight naked and blind nestlings being born after a gestation period of nineteen to twenty days. The eyes open at ten to thirteen days, the life span being between two and four years.

The Harvest Mouse is the mouse of the cornfields, reed-beds, bushy undergrowth and young woodlands. It is an attractive mouse with a reddish-brown outer coat and is one of the smallest European mice, being just over two inches long with a tail only slightly shorter. It often weighs only a quarter of an ounce, even when fully grown. It is distributed widely across the centre of Europe, from England to Russia but is not found in Scandinavia or in the most southern parts of Europe. In Britain it is becoming localized, even rare, due to modern farming practices which use combine harvesters and insecticide sprays.

The world-wide commensals of man are the Black or Ship Rat (*Rattus rattus*) and the Brown or Sewer Rat (*Rattus norvegicus*) which are found doing much damage with serious consequences. The Black Rat is smaller than the Brown Rat and is largely confined to the watersides of Europe, frequenting dockland, rivers and canals. It is the usual species found on ships but lives naturally in trees in the truly wild state. The Brown Rat is usually found living wherever unprotected food and shelter are available, such as in farms, cellars, sewers, rubbish-dumps and factories.

The Common Hamster (*Cricetus cricetus*) is scattered throughout woodland and also on lower mountain slopes. Shaped rather like a guinea pig, with short legs and tail, it lives in underground colonies, forming extensive burrows. It hibernates and will store food to see it through the winter period. It is not found in Britain, but is distributed from western and central Europe across Russia to Siberia. Mating takes place from May to July, four to eighteen nestlings being born after a gestation period of three weeks. The life span is quite long, between six and ten years.

The domesticated hamster is derived from a single family of the so-called Golden Hamster (*Mesocricetus auratus*) found on Mount Aleppo in 1930. Certain authorities consider this to be the result of a chance crossing between the Common Hamster and the Chinese Hamster (*Cricetulus griceus*), a rare case of fertile interbreeding.

Other rodents of the forest temperate areas are the plump dormice, four species being found: the Oak or Forest Dormouse (*Dryomys nitedula*), Common Dormouse (*Muscardinus avellanarius*), Garden Dormouse (*Eliomys quercinus*) and Edible or Fat Dormouse (*Glis glis*). All are attractive small rodents with rounded ears and hairy tails. The three-inch-long Common Dormouse is a reddish to chestnut brown including the two-and-a-half-inch tail, with paler underparts. The chest and neck are white and the tail often has a white tip to it. Although sometimes referred to as the Hazel Dormouse it does not only occur in this type of habitat; it frequents parks, gardens, woods such as pine and beech woods, and bushy places. A rounded summer nest of grass, leaves and plant fibres is built up to six feet from the ground. In winter this dormouse goes into deep hibernation in a nest under the ground or close to it. In Britain it is unfortunately no longer common, but in continental Europe it ranges through west and central Europe and south and east to Greece, Asia Minor and Russia. It mates from May to July, the three or four naked young being born twenty-four days later. Blind at birth, their eyes open at eighteen days, the life span being up to five years.

The largest dormouse is the Edible Dormouse which resembles a small Grey Squirrel, including a bushy tail. It was introduced into Britain in 1902 by Lord Rothschild and is now established along the Chilterns. It tends to live in colonies based in a building or in woodland.

The Coypu (*Myocastor coypus*) is South American in origin but is now feral in Europe. It was introduced into the British Isles around 1930 for fur farming, but some individuals escaped and became wild. It is now widely distributed in Norfolk, well established in parts of east Suffolk, and occurs sporadically in other areas. It has a massive head with large orange-yellow incisors, and short round hairy ears with a tuft of black hair in the centre. It has a luxuriant coat and this is one of the reasons it was used in the fur industry, under the name of Nutria.

Along the reed-beds and marshes its loud low grunts can be heard; when swimming it makes 'mooing' or humming noises, and if distressed it gives a harsh, cat-like scream. Facts about breeding in the wild state are scant, but the young are very advanced and can even swim shortly after birth. They can be suckled in water, from teats along the sides of the mother's body.

*: female Wild Boar (Sus scrofa)
her striped piglets.
nowadays Brown Bears (Ursus
live mainly in mountain*

Left: *introduced into Britain from America, the* Grey Squirrel (Sciurus carolinensis) *has colonized most of the country, but it has not spread to the mainland of Europe.*

Above: Red Squirrel (Sciurus vulgaris) *in Denbighshire, Wales.*
Below: *Resembling a small squirrel, the* Edible Dormouse (Glis glis) *was considered a delicacy by the Romans.*

The Beaver (*Castor fiber*) became extinct in England in the twelfth century, but it survives in parts of Norway, Sweden, Germany, France, Poland, Russia, Siberia and northern Mongolia, in remote areas by running water. It builds dams of branches and mud across a stream thus keeping the water level of its home territory constant, ensuring that the entrance to its living quarters or lodge remains submerged. This efficient swimmer and diver feeds on roots, bark and water plants, and during winter it does not hibernate but takes these from stores it has hidden underneath the ice. It mates from February to March, the nestlings, usually two to four, being born after about nine weeks. The Beaver's life span is estimated to be up to fifty years.

The rabbits and hares used to be grouped with the true rodents or Rodentia, but are now separated into their own order Lagomorpha ('shaped like a hare') on the grounds that they have an extra pair of incisors in the upper jaw. The European Rabbit (*Oryctolagus cuniculus*) with its yellow, brown and grey hairs used to be a familiar creature of the countryside. Evidence points to its origin being Spain from where it spread throughout Europe. It was introduced into Britain in the thirteenth century by the Normans, as a food and fur animal to replace the extinct Beaver. It was extremely widespread and quite common in Britain until it was severely hit by the myxomatosis epidemic in 1953. This was first detected in Kent and the rabbit population was nearly wiped out, but survivors now show an increase although cases are still reported.

Before this terrible disease, rabbits were abundant in grassland, cultivated fields and woodlands, and also on saltmarshes, sand dunes, mountains and moorland. They usually live in warrens, which may be complicated burrow systems or just short tunnels. Stoats, cats, badgers and rats are known to share with rabbits in communal burrow systems. Since the epidemic in Britain, however, the new population of rabbits has taken to living more above ground.

The rabbit's breeding potential is infamous, but perhaps few know the facts. It breeds mainly from January to June, but sporadically at all times of the year. The gestation period is only twenty-eight days, the litter size being from three to seven, the number increasing through the season. The rabbit crops a wide range of herbage, selecting the more nutritious species of grasses and herbs. It is greatly attracted to agricultural crops, and damages cereals, root crops, pastures and young trees with the result that it is hunted by farmers.

The rabbit may be mistaken for the Brown Hare (*Lepus europaeus*) which ranges over nearly the same areas. Although rabbit-like in build it is

Left: the fur of the Wolverine (Gulo gulo) *is immensely thick and does not retain moisture.*
Right: Red Fox (Vulpes vulpes) *vixen with her cubs.*
Below: the white winter coat of the Arctic Fox (Alopex lagopus) *helps it to conceal itself in the northern snows.*

larger and more leggy and has more prominently black tips to its longer ears. It does prefer the open flat country below 2,000 feet with a preference for the neighbourhood of cultivated land. It is also found in deciduous woods, on moors and dunes. In autumn it is often a visitor to the vineyards in Europe.

On the whole the hare is solitary, being active mainly at night but quite often seen during the day. It moves very fast with a leaping gait and with its tail often stretched out. Its 'form' is made in well-sheltered long vegetation in scrub, ditches or woods. The form is a shallow cavity into which the hare's body fits, with its head pointing out towards the scraped-out earth and its hind-quarters in the deepest part.

The courtship and aggressive behaviour of hares is spectacular and proverbial, but no systematic account is available. Although usually solitary they apparently aggregate in companies and 'box', chase, and leap considerable distances, these performances usually marking the onset of the breeding season. Mating is reported to take place from January to August with up to four litters produced in a season. The young hares or leverets, fully haired and with eyes open at birth, are placed in the form by the female. Large litters are reported to be distributed between more than one form.

The rodents, lagomorphs and birds still make it possible for several small carnivores to survive; the Stoat (*Mustela erminea*), Weasel (*Mustela nivalis*), European Polecat (*Mustela putorius*), Beech or Stone Marten (*Martes foina*), Pine Marten (*Martes martes*), Eurasian Badger (*Meles meles*) and Eurasian Otter (*Lutra lutra*).

The Stoat has brown upperparts and yellow or yellowish-white underparts, the tip of its long tail being always

Left: young European Rabbit
(Oryctolagus cuniculus).
Above right: a male Red Fox (Vulpes vulpes) *brings food for the cubs; the vixen dashes out to snatch it from him.*
Right: the Wolf (Canis lupus) *is found in Eastern Europe, Italy and Spain.*

black. It has a wide range of habitats, from lowland agricultural country, marsh and woodland to moorland and mountains. It is an excellent climber but takes to trees only if necessary; usually it stays on the ground, moving along in bounds. The average distance between bounds, measured from tracks in the snow, is twenty-two inches for males and twelve inches for females.

The Stoat hunts mainly at night when it detects its prey chiefly by smell. A single rabbit will be pursued relentlessly through an apparently undisturbed colony of feeding rabbits. A rabbit has often been observed to act in an apparently panic-stricken way, lying down at the approach of a Stoat. The kill is always made by a bite at the back of the neck. This may have given rise to the false idea that Stoats and Weasels suck the blood of their prey.

The Stoat moults twice a year; in spring it moults slowly, the moult progressing from back to belly, whereas the autumn moult is swift and takes place in the opposite direction. The autumn moult in the northern colder parts of its range changes the Stoat's pelage to white, except for the black tip of its tail. This is known as the ermine condition, the Stoat now being well camouflaged for its snowy winter home.

The Weasel is smaller than the Stoat, with a relatively shorter tail which lacks the black tip. It has almost the same colouring but lacks the straight line between the brown upperparts and the white underparts. The Weasel also turns white occasionally in its northern range.

Hawks, owls and larger mammals will occasionally kill a Weasel or Stoat, but man is their most important predator. They are widely regarded as vermin, and trapped heavily by gamekeepers. Both, however, do benefit man as they destroy large numbers of rabbits, rats, mice and voles.

The slender, sleek European Mink (*Mustela lutreola*), larger than the Wea-

Left: Red Fox (Vulpes vulpes) *cubs have very round heads and greyish fur which changes to yellow-brown in the autumn.*

sel or Stoat, is uniformly brown in colour with a white chin and usually a white upper lip. Its feet are slightly webbed, indicating its aquatic habits. It lives mainly in swamp areas, along lake borders and larger rivers. Here it builds tunnels along the waterside, often between alder roots. Being an excellent swimmer, it takes mainly water animals, including fish, but will also hunt small land game and raid farm stock.

Only one litter is produced a year; mating takes place between April and the end of May, the three to six kittens being born after nine weeks gestation. The eyes open in three weeks.

The European Polecat is rather similar in size and build to the mink, but the polecat has a longer and bushier tail, the fur being longer and coarser. It inhabits lowland areas near to water, often near farm buildings and villages. It will ascend to above the tree line in the Alps in summer. It uses a hole in the ground as a den, often a disused fox or rabbit earth. The European Polecat is often called Foumart as it can produce a foul stench. (Foumart is an abbreviation of Foul Marten which in England used to be its common name.)

Above: the Northern Lynx (Felis lynx) *is distinguished by its long pointed ear-tufts and short, 5-inch tail.*
Below: the Wild Cat (Felis sylvestris) *resembles a large domestic tabby but its tail is heavier with a broader tip.*

It marks its home range with this secretion mixed with urine.

The Eurasian Badger was formerly called Brock from the Gaelic, and this name is included in many English place names such as Brockhampton, Brockhurst and Brockenhurst. The species is distributed over most of Europe and is even found on some Mediterranean islands. It appears grey at a distance due to the colour of individual hairs which are light at the base and tip, with a dark patch between, nearer to the tip. Hair colour does vary from white through a range of shades to black. The head is white with conspicuous black stripes on either side including the eye region. The eyes of the badger are small, the ears short and tipped with white. The moult lasts through nearly the whole of the summer.

Badgers have short legs but these are very powerful diggers, used to excavate their set in a variety of places. Usually the set is found in woods or copses, especially when they border pastureland. The badger will often enlarge an old rabbit warren, and emerges from this home round about sunset for its nightly wander. It may come out earlier

Bats are the only true flying mammals; other so-called flying mammals only glide. They are mainly nocturnal and all the European bats are insectivorous, usually catching their prey on the wing. Shown above is the Whiskered Bat (Myotis mystacinus) *in hibernation. On the right, a young Serotine Bat* (Eptesicus serotinus) *clinging to its mother's fur as she carries it on her hunting flight.*

or later, depending on the time of year; in midsummer it has been known to emerge before sunset. It moves with a characteristic amble or trot. It cannot climb, but can just about scramble up on to a tree stump and loves to bathe in a stream. It lives socially in its set, but each individual tends to behave independently, and quite a number do live solitary lives. The set usually has several entrances, and sometimes a fox or rabbit will share the set. On leaving the set a badger will usually hesitate and sniff the air and listen intently to make sure all is well in the area. It is a noisy animal, snorts, moans, squeals and growls being included in its repertoire.

The cubs are usually born in the first few months of the year, but they stay underground for about eight weeks. Weaning starts at about twelve weeks when the sow regurgitates semi-digested

food. The cubs remain with the sow until autumn and sometimes over winter.

The Eurasian Otter extends over most of Europe, being quite widely distributed and living near large streams and river systems. It may alternate between sea and fresh-water habitats, but numbers found in an area are never high. Some observers say that one otter to six miles of stream is quite normal. Often all that is seen of an otter is its flattish head as it swims through the water. Its long thick tapering tail acts as a rudder, the webbed feet helping with movement in its aquatic home. One tends to think that an otter is a rich dark brown, but the colour of the dense fur is variable. Cream and white have been recorded and the tone varies with the season, being darker and richer in winter after the autumn moult.

The otter frequents waters which offer good amounts of food—freshwater fishes and shrimps in addition to frogs, birds and beetles. In some coastal waters it has been taken from crab and lobster pots, caught in the act of feeding on these crustaceans. In some mountainous areas it is suggested that the otter will migrate up with fish that are on their way to spawn, and will then follow the offspring down to the coastal waters in the spring.

The otter is mainly nocturnal, lying up in its holt in burrows, drains or hollow trees during the day, although in summer it occasionally rests in woods and reed-beds. Otters, both wild and tame, have been observed whistling, although they remain silent for long periods. Other noises have been heard only from tame ones; cubs make a whickering noise and high-pitched piping. This piping seems to keep the animals in contact with each other. Contact between adults is by a short repeated whistle. Very little is known of the otter's breeding habits. Bitches appear to produce cubs throughout the year, although certain authorities say the peak is in spring, yet others say autumn and winter.

The grassy steppes which extend across Europe and Asia, from the Hungarian plains to the north of China are a special environment. It is relatively flat with few mountains and is generally open with only occasional trees growing. This treeless area might seem to be the result of destruction by man, but in fact it is due to the climate. Because of its distance from the sea, it has become very dry with little rainfall and there is a great difference between summer and winter conditions. On the outer limits are forests which give way to prairie, and this gradually becomes arid desert in the central corridor.

Despite the harshness of the environment, and the adaptations necessary for survival in these prairies, many animals do survive. As in many regions, the rodents are best represented. European ground squirrels or sousliks are dwellers in these dry open plains of central and Eastern Europe where the soil is loamy and rich in lime. There are two species of sousliks indigenous to Europe, the European Souslik (*Citellus citellus*) and the Spotted Souslik (*Citellus suslicus*). Both frequent the same areas, the latter being identified by its spots. They live

Left and above: the stiff spiny hairs on the back of the European Hedgehog (Erinaceus europaeus) *protect it against most predators. The spines are soft at birth but soon harden.*
Right: the Eurasian Badger (Meles meles) *is a nocturnal animal and is distributed over most of Europe.*

in large colonies in which most individuals have their own living quarters. They nest at the end of their tunnels deep in the ground and when they leave their burrows to feed, mainly at night, they check that all is safe by sitting up like a marmot and listening, looking and sniffing. If frightened they give a loud, abrupt whistle but if all is clear then they move out to feed on the vegetation. They have very voracious appetites in spring and early summer, but in very hot or very cold conditions sousliks become torpid, and in very cold weather they do not eat at all.

Another excavator of the steppe is the Bobac Marmot (*Marmota bobak*), which digs its burrows in flat and hilly country in dry ground with the entrance usually facing south. Its range has diminished in the last fifty years and now extends from southern Poland to Trans-baikal and Mongolia in Asia.

The Lesser Mole Rat (*Spalax leucodon*), inhabiting eastern Europe, Hungary, the Balkans, southern Poland and southern Russia, tunnels with its broad head on workable soil including farmland. The head and fore-quarters are yellow-brown with reddish outer fur. The back is greyish, becoming slate-grey below. The completely fossorial existence it leads has led to its eyes being completely covered by skin, with no visible external ear, and enormous rodent incisors that project outside the mouth even when it is closed. Other details of its life are little known. It eats roots, bark and occasional worms and insects. Although a rodent it resembles a large mole, being up to ten inches long including the tail, as compared with the six-inch mole.

For animals to become adapted to life in high mountains, a whole series of problems has to be overcome. The human race is not very good at acclimatizing to heights, as was clearly illustrated by the 1968 Mexican Olympic Games. When men climb high mountains oxygen masks have to be worn. This is because of the lowered oxygen pressure so that less oxygen is available for the breathing mechanism. However, in most mammalian groups that are found dwelling on the mountains, acclimatization has been quite rapid. The oxygen factor is not the only limiting factor; for many it is the great temperature changes which occur, both seasonal and daily, that limit an animal's chance of surviving. However, just as certain species migrate long distances to the warmer southern areas to escape the cold northern winters, so certain mountain species move down to the foothills and the warmth at colder times of the year. An important limiting factor is, of course, food. Above the tree line, grass-eaters can only survive if they take

alpine plants, and higher still they have to have good balance to reach the shoots in crevices and on the edges of glaciers. During the winter when food is extremely difficult to come by, some hibernate by burrowing or come down to a food source below the tree line.

In Europe there are few mountain dwelling species when compared with African or American numbers. Again the rodents have succeeded in making a firm stronghold, the Alpine Marmot (*Marmota marmota*) being perhaps the most characteristic. It is found in the higher regions between 3,000 and 8,000 feet in the Massif Central, the Balkans, the Appennines, the Abruzzi and the Caucasus. It has been introduced into the Pyrenees, parts of the eastern Alps and the Carpathians with success. This quite large thick-set rodent with a broad rounded head and short legs walks with an amusing waddle as it is slightly bow-legged. Sometimes it breaks into a short gallop but is not often seen leaping. As with the other marmots, it lives in colonies with guard marmots at the entrances sitting alert and watchful on hind legs. The alarm is given by a high-pitched whistling yap and the marmots speedily disappear down the tunnels.

The Alpine Marmot is diurnal, often abroad in bright sunshine and preferring not to wander in the hours of darkness. It avoids the harshness of winter by hibernating in a nest chamber at the end of a long and complicated burrow which reaches deep into the ground. During the summer months, the marmot colony inhabits smaller, less complicated burrows and the creatures are often to be seen stretched out sunbathing or enjoying a dustbath near the tunnel's entrance.

The Snow Vole (*Microtus nivalis*) loves sunny hills and mountain-sides in the French and Swiss Alps, the Pyre-

nees, the Massif Central, the Balkans, the Apennines, the Abruzzi and the Caucasus. It is found above 13,000 feet in the Swiss and French Alps. This light grey rodent with a long tail and long white whiskers appears with regularity as soon as the sun comes out. It can climb with ease and jumps crevices and rocks with agility. It is bold enough to enter mountain huts when climbers are resting there. It runs high on its legs holding its tail up in the air. Although it digs a fairly complicated tunnel system with several openings, nest-chambers and store-chambers, the system is usually just beneath the surface with

Left: the Grey Squirrel (Sciurus carolinensis) *has been introduced into Britain from America.*
Above: the eyes of the Common Eurasian Mole (Talpa europaea) *are only 1 millimetre in diameter.*
Right: the aggressive little Common Shrew (Sorex araneus) *measures only 3 inches long.*

little depth. During the breeding season, which lasts from May to August, the voles make continuous chattering noises unlike the abrupt penetrating notes which they make at other times. After three weeks gestation usually four to seven blind and naked young are born in a nest of hay and stalks. In summer they can be found in mountain pastures, especially in the region of Alpine roses where they take grass, shoots, bark and seeds.

The symbol of the Alps is seen by many as the Chamois (*Rupicapra rupicapra*), which is the agile goat-like creature of steep mountain slopes. Originally it was confined to mountain woodlands with steep, rocky slopes, inhabiting both coniferous and deciduous woods and those that were mixed. Today it frequents the steep barren rocky slopes just above the tree line. Although one tends to think of a purely Alpine distribution, the Chamois is found in several areas in Europe; in Spain it is found in the Cantabrians, and in the Pyrenees, Carpathians, Balkans, and Appennines, as well as the French, Swiss and Italian Alps. It has been introduced successfully into the Black Forest in Germany and the Vosges mountains. This quick and extremely agile creature lives in small flocks, under the leadership of an old ewe with the males taking their place in the rear. The call is goat-like, a bleat which is used for communicating within the herd. When in danger a hissing or whistling noise is made.

The Chamois changes colour depending on the season; in summer the coat is mud brown to lemon-yellow, with a dark stripe reaching from the top of the head down each side to the corner of the mouth. The tail is dark brown above and paler below. In winter the coat is

Below: the Harvest Mouse (Micromys minutus) *is one of the smallest of European mice. It makes use of its prehensile tail in climbing stalks.*
Right: a Wood or Long-tailed Field Mouse (Apodemus sylvaticus) *being attacked by a typically aggressive Robin* (Erithacus rubecula).
Far right: a female Wood Mouse with her two half-grown young.

dark brown to brownish-black, white below and longer haired to meet the more severe conditions.

The rut lasts from November to December with kids being born the following May after twenty-one weeks of gestation. The kids at birth have thick pale reddish coats and soon become as agile and skilful on the rocky slopes and precipitous turf as their parents. The Chamois survives on grass, buds and mountain berries, rarely moving too far above the tree line.

Another goat-like creature is the Ibex (*Capra ibex*) which is often called the Wild Goat. It was once as numerous as the Chamois in the Alps but numbers are seriously depleted and it is now very rare. It has been reintroduced into several Alpine areas but is indigenous to Gran Paradiso National Park in Italy. It is larger and more stockily built than the Chamois, being four feet long compared with the Chamois' three and a half feet. Its horns are stronger, with knobs and furrows, and often reach three feet in the male. The billy also has a 'goatee' under his chin. The rut in this bovid occurs in January and the woolly coated kid is usually born in June after five months development. The young and juveniles stay with the females and together live a separate herd existence from the billies except during the rutting period. The very old females tend to lead solitary lives.

Stretching as a broad belt across the Northern Hemisphere is a region with a low average temperature where, for a

large part of the year, snow is on the ground. The natural vegetation is of the evergreen coniferous type. In Europe this region, known as the taiga, covers most of Norway and Sweden and northern Russia. The very short winter days and very long summer days affect the fauna found in this region. The average temperature of this cool temperate zone is below 40°F (4·5°C) but there is a short and surprisingly warm summer when temperatures often go over 70°F. The forests are sparsely inhabited; hunters and trappers were the only occupants for centuries, but now in the economy of a civilized world, logging and timber-working industries take place, so wild animals are again threatened. All the mammals of this zone are protected from the severe conditions by a thick coat of fur, but this means that they are hunted for their valuable pelage.

The typical image of the Wolf (*Canis lupus*) is that of a small pack wearing down its frightened prey across some snowy forest scene. The Wolf, looking like a large Alsatian dog, does inhabit the taiga but it is also found in other remote mountain forests including those in Spain, Italy, Asia Minor, Russia and Mongolia. The Wolf used to occur in England but disappeared in the sixteenth century and followed the same course in Scotland and Ireland over the next two centuries. It was very common in Europe until the last century but now fights for survival.

The Wolf is mainly nocturnal and

usually a pack is made up of a family unit, but in winter when food is scarce small units may join together as hunting packs, usually numbering less than thirty individuals. A den is made in bushes, in between roots of trees or under a rock; or a hole of some other species is enlarged. During the months from October to December the Wolf enjoys making deep howlings, and it seems that this noise is used as a mating cry. Usually between four and six young are born between February and April, their eyes opening after fourteen days.

The Wolverine or Glutton (*Gulo gulo*) has its character summed up in its name! It is hated for its greed but coveted for its fur and this has led to the Wolverine's disappearance from most of the taiga. It survives only in remote areas today, usually mountain woodland and often near marshy areas. This short-legged but large and hairy carnivore with its short, bushy tail is the most versatile of predators. It will feed on fish, frogs, voles, hares and deer, even stealing bait skilfully from traps. It is active by day as well as by night but in the summer it tends to hunt mainly at night. It is seen singly or as a mating pair, its lair being found in a shallow scrape on a rocky slope, in thick bushes, or between rocks.

A largish carnivore, the Northern Lynx (*Felis lynx*) has long legs, tufted ears, and a short tail. It is spotted on its sides and legs but not on its back. The short tail has a broad black tip. Together with the European Wild Cat, the Northern Lynx was the wild feline of the primeval forests of Europe. Now they are both exceedingly rare. The lynx can be found in large numbers only in the USSR and even there it has become rare

a tree and springing on the surprised animal. It lives a solitary life except during the mating season. The lair or den may be a hollow tree, a hole under a rock cave or in dense thickets. The male departs after the breeding season, and the female copes with the young which stay with her for a long time.

Always arousing great interest are the Norway Lemmings (*Lemmus lemmus*) and their so-called 'mass-suicides' en-

Left: the Blue or Mountain or Varying Hare (Lepus timidus).
Above: young European Rabbit (Oryctolagus cuniculus).
Above right: European Hare (Lepus europaeus).
Right: the Fat Dormouse (Glis glis) *becomes very fat before hibernating.*

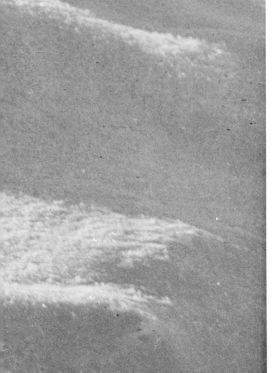

in the deciduous forest, preferring the less-populated taiga zone. In Sweden and Norway a century ago, it was fairly common in the central and southern parts of the countries. The lynx was driven north under the pressure from man and his firearms. Recently it has gained some lost ground as it has been strictly protected since 1928. In Poland it is pleasing to hear that the lynx population appears to be greater than in the 1930s. It is protected in Yugoslavia but elsewhere in Europe it is virtually extinct, although a few survive in the Carpathians and in Spain.

The Northern Lynx, with its tufted ears, is an elegant cat, and is seen usually after sunset at dusk. At certain times it can be caught sunbathing. It rarely runs, but moves in a characteristic creep at a walking pace. It frequently catches its prey by ambush, waiting in

countered in years of over-population. This small, thick-set rodent has been the focus of many research enquiries to find out the reasons why mass emigrations occur in a seawards direction, resulting in death. Recent work by ecologists from Sweden, Finland, Germany and Russia have shown that reality and myth are not equal. The lemmings winter in Scandinavia's high mountains under the snow and feed on mosses, grasses and other vegetation which stays alive under the blanket of snow. Litters are produced in this warm hidden home and, with the spring, the lemmings emerge and migrate down the mountain slopes to their summer range. This may take them only a 1,000 feet lower in the alpine zone or may end in the forest belt below. Here they settle down and produce a second litter of the year. In September the journey upwards begins to their winter quarters.

The lemmings only attract the attention of the public when there is a 'population explosion'. This does not occur every year; most years the lemmings are not numerous enough for this to happen and as they travel singly during their migratory movements under the cover of darkness they are little noticed. When a combination of factors give favourable conditions a population explosion is produced and as the large numbers move, they are noticed even at night. Most people think that mass drownings happen at the end of the migration, whereas evidence points to the fact that lemmings drown where they have become accidentally overcrowded into natural traps such as the meeting of two streams. Lemmings are good swimmers, but only enter water with hesitation.

They do, however, cross streams and lakes in large numbers, but these movements cannot be said to be 'suicidal'. Swedish naturalist Kai Curry-Lindahl carried out observations during a peak population in 1960 with some interesting results. He suggests that when enough lemmings are crowded together in a small place they will begin to behave in a way that in human terms might be described as mass panic.

Tension is built up as new arrivals increase numbers, and finally the animals are triggered off into a mass movement. This takes them in a steady direction but with no particular destination. If they encounter water, their usual reluctance to enter it is no doubt less strong than the tensions propelling the mass forwards. A narrow stretch of water is easily crossed and the migration continues, but in too wide a stretch such as a fjord or arm of the sea they will all drown.

The Blue or Mountain Hare (*Lepus timidus*) is a stockier hare than the Brown Hare of the European temperate regions. It ranges northern Europe successfully having adapted well to the adverse conditions. It is usually thought of as ranging the snowy wastes of the Arctic but it is also found in northern Scotland, Ireland, Iceland and the Alps. It has shorter ears than the Brown Hare, resulting in less heat loss from the surface area. In the summer the coat is a flat brown colour with the tips of the fur giving a grey or blue appearance. The autumn moult changes the colour to white to merge with the winter snows. It is interesting to note that the Blue Hare from Norway was introduced into the Faeroe Islands in 1820 and the stock at first turned white in winter but lost this ability some forty-five years later.

The Blue Hare is rather bold and confident, but it is not as fast as the Brown Hare, and it moves in a less zig-zag manner. It also appears to be more social than the Brown Hare, as concentration of numbers occurs in the mating season and also during the winter. This herding has been observed in other northern animals. Breeding is at its peak in March tailing off in July. Usually a litter of two to three individuals is produced, and there may be three litters a year.

Left: the Eurasian Otter (Lutra lutra) *lives near large streams and rivers, but sometimes travels far from water.*
Above right: Weasel (Mustela nivalis). *This small carnivore is tolerated by man since it preys on rats and mice.*
Right: the Stoat (Mustela erminea) *is larger than the Weasel and has a longer tail with a black tip.*

The tundra of the Arctic is even poorer in mammals than the taiga. Here the winters are long and very cold and the summers are short and sharp. The natural vegetation that survives these harsh conditions consists of certain mosses and lichens, with stunted bushes and trees, such as willow and Arctic birch, near the forest limit.

Spread over the whole of the Arctic Circle, migrating further south into Norway, Sweden and Finland in the winter, the Arctic Fox (*Alopex lagopus*) is perhaps the most characteristic mammal of the region. These attractive foxes seem to withstand extremely low temperatures, having been found in the most inhospitable places. They have been reported over 300 miles from the edge of the inland sea, wandering over the ice. They cross from one piece of land to another by drifting on the pack ice, and swimming from one ice floe to another.

In the summer the fox has a wide variety of food available—birds, lemmings, hares and voles. Records show that when there is a population increase in the lemming a similar rise is noticed in the Arctic Fox. However, during lean times the fox will eat berries and often Reindeer calves, as well as scavenging on the seashore for any shellfish or carrion.

During the harsh winter Arctic Foxes are to be seen stalking the Polar Bear, but not with the idea of attacking this powerful carnivore. The fox is extremely cunning, keeping his distance and taking care not to get too close to the bear's forceful lashing paws. The fox waits for the bear to find and kill a seal; the bear usually eats only the blubber, so the fox, often in a pair or threesome, can enjoy the leftovers of meat and entrails.

The breeding season begins in March, with the female giving birth to a litter of

Top left: Norway Lemming (Lemmus lemmus) *and young. This lemming is aggressive and will defend itself by biting fiercely. Its habitat is mountainous areas around the Arctic.*
Left: the Alpine Marmot (Marmota marmota) *is very alert and sits up on its haunches to look around and keep watch outside its burrow.*

Top left: the Common Field Mouse or Wood Mouse (Apodemus sylvaticus) *is a nocturnal rodent which eats vegetable matter, insects, and occasionally young birds and frogs.*
Left: the Pine Marten (Martes martes) *often rears its young in a hollow tree trunk.*
Above: the Common Dormouse (Muscardinus avellanarius) *is becoming very rare in Britain.*

between five and eight young in a den made in a rock fissure or in a hole dug by the pair. Often an extensive burrow system may be built. The male does not desert the family but helps with capturing food. Mating may occur quite soon after the first litter, and a second litter is born in July or August. The family remains as a unit until the Autumn.

The Reindeer (*Rangifer tarandus*) is another characteristic species of the tundra and is found in both wild and domesticated forms in its northern homeland. In its wild form it is present only in southern Norway and in Lapland. Most are migratory, moving southwards for as much as 190 miles (300 kilometres) often in very large herds. The Reindeer is a gregarious species, although in summer the mature males are solitary. They join the herds, made up of between twenty and thirty females and young, only during the period of rut which occurs in September and October. During the following months, rut herds may join together, the males following at a distance. The antlers are shed after the rut and do not start to grow again until the following April. The females are unique among the deer family in also having antlers, although these are much smaller than those of the males. The females, however, do not shed theirs until after the birth of their young in May or June. Usually a single calf is born and is weaned by the following rut period.

The powerful Polar Bear (*Thalarctos maritimus*) is an enormous carnivore and perhaps the most dangerous of all flesh-eaters. The average weight of the males is 900 pounds, the females usually being noticeably smaller averaging around 700 pounds. The Polar Bear has a long head with a Roman nose, a long neck, powerful limbs and broad feet with hairy soles (to protect the feet from frost-bite). It is amazing that this mammal can for months on end amble about the frozen ice, swimming happily for days in freezing waters. Polar Bears wander their formidable environment the whole of the year, even when the temperature drops as low as $-65°F$. Up to date little work has been done on the life of the Polar Bear, due to the difficulties of tracking it over such a vast range. Unlike other mammals, it does not seem to have a home range; its food sources are never in the same place and are often quite wide apart. Shelter is usually found by the female when she excavates a den in the snow so that she can give birth and rear her young. Usually two cubs are born and they will stay in the den with the mother for as long as five months, the cubs being suckled by the mother who survives on her reserve store of fat. The small family emerges in March or April, the mother leading her cubs down to the sea ice. The young explore this new environment eagerly, have playful scuffles and wrestling matches, but they still watch their mother, learning from her actions.

Except when the mother is with her young, or when male and female are paired during the mating season, the Polar Bear is a solitary individual. Occasionally it is seen in larger numbers when there is a large supply of food available; once as many as forty-two individuals were counted near a dead whale. The ponderous shuffle of the Polar Bear is apt to be misleading for it covers ground more rapidly than one might expect. If pressed the bear will

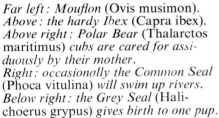

Far left: Mouflon (Ovis musimon).
Above: the hardy Ibex (Capra ibex).
Above right: Polar Bear (Thalarctos maritimus) *cubs are cared for assiduously by their mother.*
Right: occasionally the Common Seal (Phoca vitulina) *will swim up rivers.*
Below right: the Grey Seal (Halichoerus grypus) *gives birth to one pup.*

break into a lolloping gallop at an estimated speed of up to twenty-five miles per hour. The bear does not slow down to enter water but, hardly breaking its stride, goes in with a huge belly flop. It is a strong swimmer, and underwater observation shows that normally the bear only uses its fore legs for movement, using the hind legs as rudders.

In the waters around Europe's shores are some outstanding aquatic mammals—the seals, whales and dolphins. The torpedo-shaped seals have fin-shaped legs which cannot be turned forwards, and so are useless for locomotion on land. Their dense fur keeps them warm when out of water and gives them a certain resistance to their watery surroundings when swimming. There are seven species to be encountered

Far left: the Bank Vole (Clethrionomys glareolus) *inhabits hedges, bushes and fringes of woods.*
Centre left: the Stoat (Mustela erminea) *is a mainly nocturnal animal.*
Left: the Water Vole (Arvicola amphibius) *inhabits small streams, brooks and stagnant water.*
Below: the Weasal (Mustela nivalis) *is the smallest European carnivore.*

around the shores of Europe. The Common Seal (*Phoca vitulina*) is often seen along coasts having low rocks and sandbanks. Occasionally it will swim a long way up big rivers and astound city dwellers. It usually lives in large herds which break up into smaller units in the winter. It is similar to the Ringed Seal (*Pusa hispida*) but lacks the ring-shaped spots or stripes. The Grey Seal (*Halichoerus grypus*) is larger with a more pointed head and coarser spotting. The Harp Seal (*Pagophilus groenlandicus*) has a large black 'saddle' on its back.

Whales are in great danger of extinction over all the oceans of the world, but one may still be lucky enough to see a whale spouting from its blow-hole. The reason for this somewhat rapid decline of the world's largest living animals is because these marine monsters provide commodities which man wants and will kill to obtain—furs, oils and meat. Because they are slow breeding several of the whales are in great danger of extinction within this century.

Two groups of whales are represented in European waters: the toothed whales which include Sowerby's Whale (*Mesoplodon bidens*), the Bottlenose Whale (*Hyperoodon ampullatus*), the dolphins (family Delphinidae) and the Porpoise (*Phocaena phocoena*); and the baleen whales which include the Rorqual (*Balaenoptera physalus*), the Blue Whale (*Balaenoptera musculus*) and the Humpback Whale (*Megaptera novaeangliae*).

The dolphins are the well-loved creatures of the Mediterranean waters. Here these sleek acrobats of the sea can be seen exploding from the waters as they dive and play in their school. They all appear to join together in their play behaviour, one moment leaping in the air, the next bunching and rolling together or 'marching' in formation. The dolphin motif was often used on Greek and Roman coins, and can also be seen on vases and pots, and on mosaics and frescos from the Greek palaces.

Polar Bears (Thalarctos maritimus) *can dive for up to 2 minutes but do not often enter the water in winter. They eat seals and occasionally fish.*

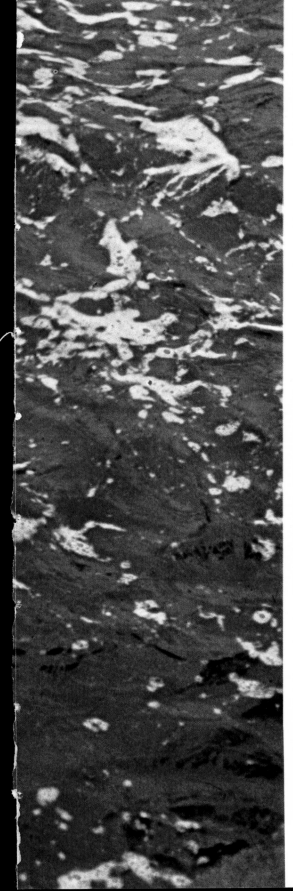

Left: the lower jaw of the Bottle-nosed Dolphin (Truncatus truncatus) *is longer than the upper.*
Above: baby Grey Seal (Halichoerus grypus).
Below: the Otter (Lutra lutra) *is a strong, graceful swimmer.*

REPTILES Reg Lanworn

When one compares the reptile and amphibian populations of Europe with those of the other continents one soon becomes aware of certain discrepancies. The pythons and the anacondas which grow to thirty feet or more are absent, the only relatives of these snakes in Europe being the sand or earth boas which scarcely attain three feet in length.

There are no crocodiles or alligators of any kind, and no ten-foot long lizards such as are found in South-east Asia. Europe has no giant reptiles alive today. In prehistoric times the great lumbering giants of more than forty feet long roamed through Europe, and their fossilized skeleton remains can be seen in many museums. Today the largest European snakes rarely exceed six feet and the largest European lizard, which incidentally is legless, reaches three feet in length. There are few dangerously venomous snakes in Europe capable of inflicting fatal bites. Those that there are are quite small, and rarely exceed thirty inches.

But with all these limitations, Europe today still possesses a very fine collection of reptile wildlife, which extends from near the Arctic Circle to the shores of the Mediterranean.

Reptiles and amphibians predominate in the southern parts of the continent, becoming scarcer and finally absent in the furthest north. The northern limits are about latitude 65° N. for reptiles, with a few of the amphibians reaching 68° N. It is astounding to find the Common Frog (*Rana temporaria*) living as far north as Cape North in Norway. It is at least safe in this part from the Grass Snake (*Natrix natrix*) which largely feeds upon frogs. The Grass Snake and the Adder or Northern Viper (*Vipera berus*) appear to live further north than any other snake, but these northern specimens are not usually so brightly coloured as others found further south. Also in these northern parts are races of melanistic snakes and lizards. In these black individuals the black zig-zag of the

Female Grass Snake (Natrix natrix) *with her eggs. Incubation takes 10 weeks.*

Above: the attractive Eyed Lizard (Lacerta lepida) *is aggressive by nature and can give a painful bite.*
Below left: fights between male Green Lizards (Lacerta viridis) *are not uncommon.*
Below right: both the Common Lizard and the Slowworm (Anguis fragilis) *give birth to live young in the summer.*
Right: the Adder (Vipera berus) *is common from the north of Europe to the South of France.*

*Above and below: the Sand Lizard (*Lacerta agilis*) has a conical head and blunt snout. It favours sandy or well drained country.*

Adder or the yellow collar of the Grass Snake cannot be discerned.

There are two species of lizards which live in the far north, the Common or Viviparous Lizard (*Lacerta vivipara*) and the Slowworm (*Anguis fragilis*). Both these lizards are numerous, and are often to be seen on commons and dampish woodland clearings in Britain on sunny days.

The Common Lizard is a small brownish creature of six inches, with a darkish stripe down its back. On its sides it has fine spotting but when the sun shines on it, it takes on a metallic appearance, similar to the Slowworm. The Slowworm itself is a legless lizard, about the thickness of a pencil and reaching approximately eighteen inches in length. It is harmless, feeding mainly on small slugs and earthworms. Its general colour is bronze when adult, but there are varieties which have blue patches suffused into the bronze, producing a most attractive effect. Both the Common Lizard and the Slowworm give birth to live young in the summer, the former lizard producing about a dozen, the Slowworm up to twenty or more in one litter. Birds reduce these families considerably as they appear to be very acceptable morsels and look worm-like at birth.

In addition to the Slowworm there are other legless lizards in Europe, but only in the warmer areas. The outstanding example is the Glass Snake or Scheltopusik (*Ophisaurus apodus*) which is the largest of this group of lizards. It is found in the Levant countries and around the arid areas of the Black Sea. The Scheltopusik is unlike any other lizard found in Europe. It has a skin of hard bony rhomboid scales which are keeled and ribbed. The head and body are continuous with no distinct neck, and down each side there is a deep groove. This ends at the beginning of the tail with minute remains of vestigial limbs.

The overall colour is dullish brown or grey but occasionally brick red specimens are found which are probably old animals. A few years ago, Glass Snakes were often seen in the stony vineyards feeding on large grasshoppers or small rodents. Nowadays they are rarely seen in these surroundings, perhaps due to modern cultivation methods and insect control, though they still appear to survive in less cultivated terrain where they feed on insects, mice, and snails whose shells they crush with their powerful jaws.

The most insignificant member of the legless lizard group is the Worm Lizard (*Blanus cinereus*). This is seldom encountered as it is usually only found under stones. The very few I have seen alive were found in cemeteries in Spain, but they are likely to be more numerous than at first appears to be the case. To the unobservant, they look like ordinary earthworms, but a closer examination shows a little blunt snout, and a tiny groove down each side of the body as in the Glass Snake. The body is covered with finely scaled skin which forms rings, and which varies in colour from a flesh colour to shades of brown.

Probably the lizards more often

observed than any others are the Wall Lizards (*Lacerta muralis*) and their allied species the Ruin Lizards (*Lacerta sicula*). They may be seen from Holland to the coasts of the Mediterranean, and even beyond, as many of the small rocky islets in these warm waters have their own individual colonies forming varieties or races of these agile little lizards. Few grow larger than eight inches long and their requirements are modest, their habitat being cliffs or rocks with numerous cracks and crevices, or even ruins of old buildings, preferably situated in a sunny southern aspect. The area must also have a good insect population. As with all the lacertids the Wall and the Ruin Lizards have a slender streamlined body, and a long thin tapering tail which is often twice the length of the body. They are covered with keeled or slightly keeled small scales.

They are fascinating animals to watch, but one clumsy move on the observer's part and they disappear into their retreat which is usually an inaccessible hole in a rock. The most successful method of capture is one I saw being practised by a small group of Italian children. They used a horsehair noose on the end of a thin cane, which they gently lowered over the unsuspecting sunbathing lizard. With a sudden jerk the lizard was caught like a small fish on an angler's line.

The Sand Lizard (*Lacerta agilis*) is slightly larger and a little more robust than the previous lizards, and has a stubby head and less pointed nose than the Wall Lizards. It favours sandy or well drained country but not hot arid parts. In Britain it is found on or near sand dunes in the south, and in France I have seen it in the Forest of Fontainbleau near Paris—two very different terrains.

They have a small range of colours from brown to green, but most specimens have a broad stripe down the back consisting of squares of alternating light and dark shades of brown, the dark squares having a white spot in the middle. When in breeding condition the males become greenish on the back and

Above: the Fire Salamander (Salamandra salamandra) *is well protected from enemies by poisonous secretions. Below: attempts to introduce the Green Lizard* (Lacerta viridis) *into Britain have failed, but it is common in France.*

underside. The females are usually cream underneath, with or without black spots.

The Green Lizard (*Lacerta viridis*) is more widespread than the Sand Lizard. It is not found in Britain. There have been many attempts to introduce it into the warmer counties of England, but the small colonies have died out after a few years. It is still to be seen on the north coast of Jersey and is fairly numerous along the Atlantic coast of France as far as the Pyrenees, in Italy and through to the Balkans. There were colonies around the lower Rhine but it has become very rare in Germany. The Green Lizard is one of the most beautiful of its kind. It grows to about a foot long, and is very sleek. Its body is sometimes emerald green, sometimes a lime green with black spotting. Some specimens have a blue throat which at one time was thought to be a male characteristic, but in Jersey the writer

found that both sexes of the lizards often had blue throats. They appear to prefer to live in open scrubby country where they can dig under dry roots or stony outcrops. It is in these hideouts that the female deposits her dozen or more eggs which eventually hatch into tiny brown lizards, so unlike the beautiful green adults they will become.

South of the Pyrennees the Eyed Lizard (*Lacerta lepida*) takes the place of the Green Lizard. It is also known in the coastal belt near Perpignan, and in Portugal. Whereas the Green Lizard is a nervous creature, rushing headlong away from any intruders, the large male Eyed Lizard has a much more aggressive attitude. He will not attempt to rush away if he is cornered, but will attack and if given the opportunity will give a painful bite.

In appearance this is an attractive lizard, attaining a length of eighteen inches, with an overall colour of pale olive, patterned with a dark net-like tracery and large blue eye markings along its sides. Eyed Lizards eat most insects, small rodents and fledgelings, even raiding birds' nests in low trees. According to reports from some of the Portuguese gamekeepers, they prove a menace on many estates where game birds are reared, by breaking the eggs or eating the young, and for this reason they are sometimes dealt with as pests.

The most fascinating of all Europe's lizards is the Common Chameleon (*Chamaeleo chamaeleon*). There is only one species in this continent and it is only found in a few restricted areas in Spain, Portugal, Cyprus and Crete. Its shape is well known but its numerous distinctive features never fail to intrigue the onlooker. The eyes set in turrets giving the creature nearly 360 degrees of vision, the parrot-like feet with opposable fingers giving a near perfect grip on twigs, and its great body-length tongue which it projects out at lightning speed to catch its food. Then there is its ability to change colour in a matter of seconds.

The European chameleon lays its eggs during the summer in a deep hole in dry earth and the eggs hatch in the following spring. Most of its life is spent in trees and shrubs feeding on winged insects, but its numbers are dwindling. Specimens are often collected and disposed of as pets, for which they are totally unsuitable. They are probably

Left: Alpine Newts (Triturus alpestris). *Below left: the Edible Frog* (Rana esculenta) *occurs over most of Europe. Below: the Viviparous Lizard* (Lacerta vivipara) *bears its young alive.*

the most difficult lizard one can try to keep, usually dying after a few months.

In Corfu and other Greek Islands the Starred Agama Lizard (*Agama stellio*) makes its home. This belongs to a genus of lizards which are numerous in Africa and Asia. They also are colour-change artists. Some species are even able to take on brighter and more vivid colours than the chameleon, but the European species can only take on various shades of light and dark. On sand it takes on a sandy colour, on dark rocks it becomes grey or near black, with various spots remaining unchanged.

The geckos are able to change from

Left: the Slowworm (Anguis fragilis) *gives birth to up to 20 young in one litter.*
Below: the Common Chameleon (Chamaeleo chamaeleon) *has nearly 360 degrees of vision.*
Right: Wall Lizard (Lacerta muralis).

light to dark shades, and to see these lizards running from a dark corner across a white ceiling is a remarkable experience. They have pads of laminae or very fine ridges on their feet which enable them to travel on any surface, even upside down. The Moorish Gecko (*Tarentola mauritanica*) is the largest, about six inches in length, and is found along the Mediterranean coastal regions. It is abundant in the cork forests and often takes up residence in the older villas. It lives in wall crevices during the day, and goes stalking flies and mosquitos at night like a hunter. Very slowly it approaches to within inches of a winged insect, then suddenly pounces to get to grips. Its adversary is sometimes a large moth, and then both contestants fall to the floor. The gecko nearly always wins, and gains a meal.

There are six species of skinks (genus *Chalcides*) with several sub-species. These burrowing lizards generally have plumpish bodies with a highly polished appearance and very short legs. They have small recesses behind the limbs into which the limbs fit when they are burrowing in hard ground, while their hard nose does most of the work. Their colour is silvery grey or pale brown, but the small hexagonal scales give a highly polished linen effect. Sometimes they lie above ground in the sun, but they quickly burrow down again if disturbed. They are only found in very dry sandy areas and feed mainly on beetles, but they will eat other insects. They bear live young, often several families in the same year, and it is a fascinating sight to observe the newly born infants energetically excavating within minutes of their birth.

All the lizards need an efficient escape routine to survive, as many form the staple diet of some birds and snakes. A large proportion of snakes that live in the dry areas of southern Europe feed exclusively on lizards, in much the same way as the snakes that live in the damp marshy area rely on frogs and fish as their main food.

The Grass Snakes (*Natrix natrix*) are never found very far from water, where they search out frogs from the reed beds or swim around for small fish. They may be seen on heaths, on the edge of woods and in forest clearings. When their natural food is in short supply they have even been known to raid goldfish pools in suburban gardens, remaining in the vicinity until the pool is denuded of fish.

The Barred Grass Snake (*Natrix natrix helvetica*) is fairly common in the British Isles, with the exception of

Ireland where it is absent. A few specimens in Europe have reached six feet in length, but the majority are much smaller. They are completely harmless creatures. They have large gaping mouths with rows of very fine teeth but they never bite humans even when handled. They hiss very loudly and throw their head forward as if to strike, but their mouth is closed. For their own protection they are able to feign death, and also emit an unpleasant pungent odour which has a lasting property on hands or clothing.

The Grass Snakes' general colouring is olive or grey green. Some have black flecking on the back, while others such as the Spanish varieties (*Natrix natrix astreptopha*) have series of black spots.

Below: the Aesculapian Snake (Elaphe longissima) *inhabits the south and warmer regions of central Europe. Right: colour variation among Adders* (Vipera berus)—*a black or melanistic form together with a normal form.*

Many of them, including the Barred, have a yellow collar or two yellow half-moon patches behind the head, but these are generally missing in the Corsican and Spanish specimens.

Among the snakes that are closely related to Grass Snakes are two aquatic species which spend most of their lives in stagnant ponds, or in streams; the Diced Water Snake (*Natrix tesselata*) and the Viperine Water Snake (*Natrix maura*). They are rather dull snakes. They have flat heads with the eyes situated on the top. The Diced Snake is usually the colour of mud and has square black markings on its back. Underneath it is brighter with a flushing of pink, chequered with black. The Viperine Snake has obtained its name from the black zig-zag marking down its back which is similar to that of the Common Northern Viper, but in old specimens the markings have often completely disappeared and the snake looks nothing like a viper.

Another snake which is sometimes mistaken for a viper is the Smooth Snake (*Coronella austriaca*). It is a rare snake in Britain, but is common in many other countries from the north to the south of the continent. It is either grey or grey-brown with small black spots on the head and body and a black horseshoe mark behind the head. It feeds chiefly on lizards, but it will take newly born field mice or voles. It is a belligerent little snake and will bite at the slightest provocation, but its teeth are very small and can do little damage. It gives birth to ten or more young during August.

Many of Europe's reptiles, including all the venomous snakes, are ovoviviparous or live bearing. The eggs remain in the female until the ova is fully formed, enabling the infant to leave the egg case immediately it is laid. This method of reproduction has given rise to a story which is still believed in country districts, that young snakes,

when danger is imminent, crawl down their parent's throat and remain inside the stomach until the danger has passed. This is in fact an impossible feat due to the rows of needle-like teeth in a snake's mouth.

As previously mentioned there are no large pythons or boas to be found in Europe, but the Boidae, the giant snakes family, are represented by the small constrictors of the genus *Eryx*, the sand or earth boas. They are only found in the very warm parts of the Eastern Mediterranean regions, from the Balkans to the eastern end of the Black Sea, where they live underground during the day, emerging at night to feed on lizards and mice. They have a dozen or more young, which are born alive and are similarly patterned to the adult, having irregular dark bands across a general colouring of pale fawn. As is common to all the Boidae, traces of rudimentary limbs can be seen either side of the vent; but in the sand boas these can only be noticed in the larger specimens.

Europe has many constricting snakes, the largest growing to more than six feet. This is the Aldrovandis or Four-lined Snake (*Elaphe quattuorlineatus*). The adults are fawn, generally with four lines down the whole length of the body. The young have blotches which gradually join up to produce lines as they grow older. This snake is a most docile creature, and even freshly caught specimens will rarely attempt to bite. It is because of this friendly disposition that they have been exported in large numbers from Italy and Albania to be sold as pets, with another snake that is usually docile within a short time of capture, the Aesculapian Snake (*Elaphe longissima*). This creamy brown snake derived its name from the supposition that it was installed by the Romans in their Aesculapian Temples in the spas and watering places throughout Europe. There appears to be foundation for this, as in many places the snakes are known only where these temples existed. Many of the constrictors are unfriendly snakes, and they will readily avoid humans and can become aggressive if cornered. The Dark Green Snake or Whip Snake (*Coluber viridiflavus*) is this type of snake, and in parts of Italy it is sometimes known as the Fury Snake because of its quick temper.

The most colourful of these constrictors is the small slender Leopard Snake (*Elaphe situla*) with its yellowy grey body and a single or double row of red blotches, each blotch having a black edging. Although sometimes nearly three feet in length, they are only as thick as a small finger and often remain during the day coiled up in a hole in a rock or an old building. They are only likely to be seen in the evenings, when they go searching for the small rodents upon which they feed.

The snakes so far mentioned are comparatively harmless. They all have needle-like teeth, and the larger specimens may inflict a slightly painful bite without any lasting result. The bite of the venomous snakes is much more serious, but few of the European vipers are capable of delivering a lethal bite. The Northern Viper or Adder (*Vipera berus*) is a common snake found from the north of the continent to the south of France. It rarely exceeds two feet in length, but it is thick bodied and has a very short tail. The males are grey and the females various shades of brown, but both sexes have a black zig-zag marking down the centre of the back and a 'V' marking on the neck. They may be encountered on cliff-tops or moors, or near woodlands where they feed on rodents and lizards. During early morning sun they may often be seen on tufts of heather on Scottish moors, remaining until they are nearly trodden upon before gliding away to safety. The Asp Viper (*Vipera aspis*) is quite similar in appearance to the Northern Viper. It does not always have such a pronounced black zig-zag, sometimes only black or brown patches, and it has a slightly turned up nose and a yellow end to its tail. It is found in

France, Switzerland and parts of Germany and the Tyrol. There is a story that many years ago the German military authorities were so concerned about the supposed infestation of a training ground near Metz that they offered a price per head for these vipers. After paying out the bonus at an alarming rate, it was dicovered that the snakes were being imported over the frontier from Lorraine, where they were fairly numerous.

The Long-nosed or Horned Viper (*Vipera ammodytes*) is the most dangerous of the European vipers. It grows larger than any of the others, nearly three feet long, and is inclined to be

Although largely feeding on frogs, the Grass Snake (Natrix natrix) *will also eat fish. Below, a carp is being attacked and, bottom, it has just been eaten. The Grass Snake is known throughout Europe and shows considerable colour variation. It can reach a length of nearly 6 feet.*

aggressive, often standing its ground when other snakes would have retreated. Its venom is particularly potent and many deaths have been caused by this snake. It can be recognized by a fleshy protuberance on its nose which looks like a small horn. Its colouring is similar to the previous vipers. The Long-nosed lives in rocky districts of the Tyrol and through the countries skirting the Mediterranean as far as Greece. The vipers are generally easy to identify. They all have a rather thick body, a short tail, very small scales on the top of the head, and vertical pupils. The European species all have more or less zig-zag markings. Their poison fangs are hinged teeth which become

erect when the mouth is opened, and which are hollow, similar to hypodermic needles. Behind these teeth there are always reserves, ready to take up position when the others fall out. The vipers, which have fangs in the front of the jaw, are known as the Solenoglypha group, but there are three snakes in Europe which are back-fanged or Opisthoglypha. These mildly poisonous snakes, which are not considered dangerous to man, only have very small poison fangs far in the angle of their jaws. The most impressive of these snakes is the Montpellier Snake (*Malpolon monspessulana*). This brown snake of about five or six feet is the most voracious feeder of the southern European reptiles. It will eat anything or everything that moves and is small enough. I have known it eat snakes its own size and complete families of fledgelings, giving it the appearance of a large string of sausages.

The wiry little Cat Snake (*Telescopus fallax*), another back-fanged snake, is found in the Balkans and the Greek Islands. It is pale grey with large black spots down its back. Its eyes are distinctly cat-like and give it a sinister

Top: from 4 to 15 eggs are laid by the Smooth Snake (Coronella austriaca) *and hatch immediately.*
Left: the Adder (Vipera berus) *is thick bodied with a short tail.*
Below: Diced Water Snake (Natrix tesselata).

appearance. If it is handled, care being exercised to avoid being bitten, it is difficult to imagine that the snake is not made of stiff wire, and it is quite a feat to uncoil it from the hand.

The third member of the group is only found in the south of Spain. It is known as the Cowl or Hooded Snake (*Macroprotodon cucullatus*) because of the hood-like patch on its head. On first examination it can be mistaken for the Southern Smooth Snake (*Coronella girondica*), but it is a nocturnal snake and very rarely encountered during the day, unless rocks are overturned and it is moved from its hiding place.

Rare reptiles are more often seen by accident than when searched for by naturalists, and this is very true regarding the large turtles which visit the coasts of Europe. Occasionally during the summer individual specimens of the great marine turtles are seen in European seas. I have seen large Loggerheads drifting in the Bristol Channel, and the Hawksbill and Green Turtle are seen from time to time in the Mediterranean. But they cannot be classed as European animals since they only breed around the tropic coasts. The ones seen around European shores are specimens which have strayed from warm oceans nearer to the Equator. Of recent years it has been suggested that nuclear activity in the South Pacific could have affected the turtles' navigational instincts and caused them to go off course, but most scientists disagree with this suggestion. In any case, there were always a few turtles to be seen outside their territorial waters long before the atom bombs were tested.

There are seven species of chelonians which are indigenous to Europe; three of these are terrapins which lead aquatic lives, and four are land tortoises which prefer to live in arid country.

The chelonians have proportionately larger brains than other reptiles, and they also respond to advances by humans more readily than do the snakes and lizards. It is probable that their friendliness is accounting for their near extermination. Every year there is a massive export of hundreds of thousands of land tortoises to be sold as pets in cities all over Europe. Vast areas of North Africa, where they lived in their thousands, have been completely cleared and now a similar trade is being carried out in the Balkans and other countries. The enormity of the trade can be envisaged when one is informed that up to half a million tortoises have been imported into Britain in one year.

It is sad to relate that only a few survive to the following year, more so when it is realized that they are normally long-lived creatures; there have been many cases recorded of them living fifty years or more in proper conditions. The northern climates of Europe are unsuitable for land tortoises unless they have the run of a very sheltered garden, and good conditions for hibernation. They simply cannot endure long spells of cold damp weather. It prevents them feeding sufficiently and they eventually die of inanition. The land tortoises are more docile than their aquatic relatives. Their main diet consists of succulent plants, grasses and fallen fruits. Any flesh they eat has usually been dead for some time and they are not opposed to eating a very decomposed animal or bird. They are also creatures of habit to a certain extent; they use the same shelter, under a bush or overhanging rock or even just dug into a particular dry patch of soil, for night after night. They awake and amble away from their shelter as the day becomes warm, and always return to it as the sun goes down. They often use the same track from one part of their terrain to another. In the spring, after they have emerged from hibernation the males, which can be identified by their smaller size and concave plastrons, begin to take an interest in the opposite sex. At this time they become definitely belligerent. When the male sees a female he becomes suddenly stimulated and rushes at an unbelievable speed towards her. On reaching her he stands with his shell clear of the ground and with a sudden thrust bangs his shell at the rear of his prospective bride at the same time pulling his head inside to prevent self-injury. This proceeding is repeated several times, after which he goes to her fore end making soft croaking noises. If his advances are accepted and she extends her head towards his, the pair wander to a more secluded place. Often other males appear before mating takes place and a miniature 'tank' battle ensues with the participants attacking each other head or side-on, and even resorting to biting one another. If a satisfactory pairing takes place and the five or six white eggs are laid, they hatch in about sixty days, producing perfect replicas of their parents but for their soft shells which harden as they grow.

Left: the Pond Tortoise (Emys orbicularis) *lives in marshes, ditches and streams of central and southern Europe.*
Above: the common tortoise of the Balkan countries is the Greek Tortoise (Testudo hermanni).

All the three species of European land tortoises have in the past been known as Greek Tortoises, and even the correct identification of the species can be confused by the English and scientific names. For instance, one would expect the European Tortoise (*Testudo graeca*) to be found in Greece, but in fact it is not. It is mainly found in North Africa and part of southern Spain. It is now more widely known as the Spur-thighed Tortoise because of the large horn-like tubercle on the back of its thigh.

Herman's Tortoise (*Testudo hermanni*) is now referred to as the Greek Tortoise. This is the common tortoise of the Balkan countries and a few districts of Italy. It has no large spurs on its legs, but has one on the end of its tail. The Marginated Tortoise (*Testudo marginata*) is found in parts of southern Greece and Italy. This tortoise, although similar to the previous two species, has no spurs on legs or tail.

Of recent years a darker grey or olive tortoise has been offered for sale by the pet dealers in huge quantities. It is a very active creature even in cool weather, but it very soon succumbs and dies in damp weather. This is the Horsefield's Tortoise (*Testudo horsefieldi*) whose home is on the Russian steppes across to Afghanistan, and in one or two small districts in north-eastern Europe. This tortoise has an easily recognizable feature. It has four claws on each of its feet, whereas the other species of land tortoise have five on the front feet and four on the back.

Male Common Frog (Rana temporaria). *This species begins to croak when fourteen months old.*

All the land tortoises have a high domed shell, rather stumpy short legs, and a small tail which is tucked under the shell out of sight. The terrapins on the other hand have a flatter shell, their legs are longer and more compressed with partially webbed feet, and they have a long tail which is not hidden away. The claws on the feet are strong and well curved and provide them with the ability to clamber over muddy obstacles, or tear up food too large to swallow whole. In all, the terrapins are efficient creatures. They can exist on land or water, can swim, dive and bury in the mud, and go for long periods without food. They will eat a variety of food, both animal and vegetable. In the very cold weather they sleep for long periods. Their shells are strong and provide protection for the whole body, including limbs and head. In view of their adaptability it is surprising that Europe only has the three species.

The European Terrapin or Pond Tortoise (*Emys orbicularis*) is a robust animal which lives in the marshes, ditches and streams of central and southern Europe. It often takes up residence in stagnant ponds where it can find enough of the snails, water insects and amphibians upon which it feeds. It may sometimes be seen sunning itself on a bank near water, but on the approach of an intruder it will dive into the water and swim to the bottom. It can remain submerged for hours but after these encounters it becomes inquisitive and swims up, using its head like a periscope before its main bulk comes into view as if to ascertain that the coast is clear. Their colouring gives

them protection in watery depths. The carapace or upper shell is nearly black with a few yellow spots or radiating lines, and the legs and head are similar. The under shell or plastron is yellow with a few dark lines or spots. But when they become old, all the light markings fade away, and specimens of seven inches long are nearly black. They lay their eggs, usually about ten, in holes a few inches deep which they dig in ground adjacent to water. The eggs incubate in eight to twelve weeks.

The Pond Tortoise used to be much more numerous in Europe than it is today. A hundred years ago farmers in parts of Germany often placed one in their animals' drinking troughs or in the large vats that held the pig-swill before it was used as food for the pigs. They believed that the tortoise would eat the worms and other parasites which caused sickness in the farm stock.

The other terrapins live in the warmer parts of the continent, the Spanish Terrapin (*Clemmys leprosa*) in Spain and Portugal, and the Caspian Terrapin (*Clemmys caspica rivulata*) in the Balkans. These two species are somewhat alike in shape, size and disposition. When young they are oval in shape, but as they reach maturity they take on a squarish look and their bright colouring fades. Their normal colouring is more attractive than that of the Pond Tortoise. The Spanish has a creamy brown shell with bright orange or yellow centres to each shield, and yellow striped neck and legs. The Caspian's shell is a darker colour, brown or olive and has a light-grey pencilling. The stripes on its legs are edged with black.

All the terrapins have powerful jaws which are capable of gripping a struggling fish or frog, and this tenacity is increased by the finely serrated upper jaw of the Caspian, and a beak-like projection of the lower jaw which fits into a notch in the upper jaw of the Spanish Terrapin. When hungry, terrapins can become persistent and will attack even large fish, and there are many reports of their tearing to pieces shot game birds which have dropped into open water occupied by hordes of

Above: the Edible Frog (Rana esculenta) *is mostly aquatic but will sometimes migrate overland.*
Right: European Green Tree-frog (Hyla arborea) *with its eggs.*
Below: the Marsh Frog (Rana ridibunda) *is seldom far from water.*

terrapins. This causes displeasure to the sportsman, and sometimes these incidents bring about the extermination of terrapins, a very shortsighted action.

This short chapter attempts to give an impression of the diversity of the reptiles of Europe. It is difficult to understand why many people have a feeling of abhorrence when the word 'reptile' is mentioned. It must conjure up something sinister for them. They fail to realize that a tortoise is a reptile and only associate the word with venomous snakes. Perhaps we require a new collective term for this interesting group, instead of the one which broadly means the 'creepers'.

In time they may be better understood by a greater number of people, and then get the protection necessary to prevent their numbers being reduced still further than they have been already. We cannot increase the numbers of species; let us at least try to keep the ones we still have. Once a species becomes extinct, it has gone for ever.

INDEX

References to illustrations in italic

Abramis ballerus 19
　A. brama 17
Acanthis cannabina 60
　A. flammea 60
Accentors 55
Accipiter gentilis 32
　A. nisus 32
Acerina cernua 16, *18*
Acipenser sturio 24
Acrocephalus scirpaceus 55, *61*
Adder 107, *108*, 110, *116*, 117, *119*
Aegithalos caudatus 58, *59*
Aegolius funereus 49
Agama stellata 114
Alauda arvensis 53
Alca torda 51
Alcedo atthis 40, 52
Alces alces 66
Aldrovandis 117
Alectoris rufa 38
Alopex lagopus 79, 96
Alosa alosa 24
　A. finita 24
Anarhichas spp. 13
Anas crecca 31
　A. penelope 31
　A. platyrhynchos 30, *39*
Anguilla anguilla 21
Anguis fragilis 110, *114*
Anser fabalis 30
Anthus cervinus 53
　A. trivialis 54
Aphanius fasciatus 25
　A. iberus 25
Apodemus sylvaticus 74, *90*, 97
Apus apus 52
　A. melba 52
　A. pallidus 52
Aquila chrysaëtus 32, *46*
　A. heliaca 32
Ardea cinerea 29, *36*
　A. purpurea 36
Arenaria interpres 31
Arvicola amphibius 101
Ascidiella aspersa 15
Auks 42, 50
Avocet 47
Aythya fuligula 31

Badger 67, 83, 84
Badger, Eurasian 81, 83, *86*
Balaenoptera musculus 103
　B. physalus 103
Barbels 16, 18
Barbus barbus 16
　B. meridionalis 18
Bat, Serotine *84*
　Whiskered *84*
Bear, Brown 66, *75*
　Polar 96, 98, *99*, *103*
Beaver 79
Bee-eater 52
Bison bison 64
　B. bonasus 63, *64*
Bison, European 63, *64*
Bitterling 20
Bittern 29, *39*
Blackbird 58, *59*
Blackcap 55
Blanus cinereus 110
Blennius pholis 13

Blicca bjoernka 16
Bluetail, Red-flanked 59
Bluethroat 58
Boar, Wild 64, *75*
Boas, Sand or Earth 117
Bombycilla garrulus 54
Botaurus stellaris 29, *39*
Branta ruficollis 30
Bream 16, 17, 18
Bubo bubo 50, 52
Bullfinch 60
Bullhead 16
Buntings 60
Burbot 16
Bustard, Great 42
Buteo buteo 32
　B. lagopus 32
Buzzards 32

Calandrella brachydactyla 53
Calidris alba *31*
　C. alpina 47
　C. canutus 31
Canis lupus 72, *81*, 91
Capercaillie 38
Capra ibex 90, *99*
Capreolus capreolus 64, 68, *71*
Caprimulgus europaeus 52, *52*
　C. ruficollis 52
Carduelis carduelis *49*, 60
　C. chloris 60
Carp 16, *16*, 18, 19
Caryle alcyon 52
　C. rudis 52
Castor fiber 79
Cat, European Wild 66, *72*, *83*, 93
Catfish 21, 22
Certhia brachydactyla 60
　C. familiaris *59*, 60
Cervus elaphus 64, *66*, *127*
Chaffinch 60
Chamaeleo chamaeleon 113, *114*
Chameleon, Common 113, *114*
Chamois 89, 90
Char 16, 24
Charadrius hiaticula 46
Chiffchaff 56
Chlidonias niger 50
Chub 16
Ciconia ciconia 29, *34*
Cinclus cinclus 54
Circus aeruginosus 34
　C. cyaneus 34
　C. macrourus 34
　C. pygargus 34
Citellus citellus 86
　C. suslicus 86
Clamator glandarius 51
Clangula hyemalis 32
Clemmys caspica rivulata 123
　C. leprosa 123
Clethrionomys glareolus 101
Cobitis aurata 21
　C. taenia 20
Coccothraustes coccothraustes *57*, 60
Cod *13*, 15
Coluber viridiflavus 117
Columba palumbus 51, *55*
Coot 39, *40*, 42
Coracias garrulus 52
Cormorant 28, *28*
Coronella austriaca 116, *119*
　C. girondica 120

Corvus corax *46*, 61
　C. corone 61
　C. corone corvix *44*, 61
　C. frugilegus 52, 61
　C. monedula 61
Cottus sp. 25
Coypu 75
Crakes 39
Crane 38, *38*
Creeper, Short-toed Tree 60
　Tree *59*, 60
　Wall 59
Cricetus cricetus 74
　C. griceus 74
Crow, Carrion 61
　Hooded *44*, 61
Cuckoo 51, *51*
　Great Spotted 51
Cuculus canorus 51, *51*
Curlew 39, 47
Cuttlefish, Common 25
Cyanosylvia svecica 58
Cyclopterus lumpus 25
Cygnus cygnus 29
　C. olor 29, *34*
Cyprinus carpio 16

Dace 16, 17
Dama dama 64, 68, *71*
Deer, Fallow 64, 68, *71*
　Red 64, *66*, *127*
　Roe 64, 68, *71*
Delichon urbica 53
Dendrocopus leucotos 52
　D. major 52
　D. medius *44*, 52
　D. minor 52
　D. syriacus 52
Dipper 54
Divers 27, *28*
Dogfish, Lesser Spotted *23*
Dolphins 99, 103, *105*
Dormouse 70
　Common 74
　Edible or Fat 74, 75, *77*, *93*
　Garden 74
　Hazel 74
　Oak or Forest 74
Dove, Collared 51
Dryocopus martius 52, *55*
Dryomys nitedula 74
Ducks 31, 32
Dunlin 47
Dunnock 55

Eagles 32, *46*
Eel, Common *21*
　Moray *21*
Egret, Little *27*
Egretta alba 29
　E. garzetta *27*
Eider 31
Elaphe longissima *116*, 117
　E. quattuorlineatus 117
　E. situla 117
Eliomys quercinus 74
Elk 66
Emberiza bruniceps 60
　E. citrinella 60
　E. leucocephala 60
　E. melanocephala 60
Emys orbicularis *121*, 122
Eptesicus serotinus 84

Erinaceus algirus 67
   E. europaeus 67, *86*
Erithacus rubecula 49, *58*, *90*
Eryx spp. 117
Esox lucius 16, *16*

Falco eleonorae 35
   F. peregrinus 35, *42*
   F. rusticolus 35
   F. r. islandus 35
   F. r. landicans 35
   F. subbuteo 35
   F. tinnunculus 34, *42*
   F. vespertinus 35
Falcons 34, 35, *42*
Felis lynx 72, *83*, 93
   F. silvestris 66, *72*, *83*
Ficedula albicollis 57
   F. hypoleuca 57, *61*
   F. parva 57
Fieldfare 59, *60*
Firecrest 55, 57
Flamingo 29, *32*
Flycatchers 55, 57, *61*
Fox, Arctic 79, 96
   Red 67, *68*, 79, *81*, *82*
Fratercula arctica 32, *36*, 51
Fringilla coelebs 60
Frogs 92, 115
Frog, Common 107, *122*
   Edible *112*, *123*
   Marsh *123*
Fulica atra 39, *40*
Fulmar 28
Fulmaris glacialis 28

Gadus callarius 13, *15*
Gallinago gallinago 46
Gallinula chloropus 39
Gambusia 25
Gannet 28, *29*, *36*
Garrulus glandarius 61
Gasterosteus aculeatus 18
Gavia arctica 28
   G. stellata 27
Geckos 114, 115
Geese 30
Glareola pratincola 47
Glis glis 74, *77*, *93*
Glutton 92
Goat, Wild 90
Gobius paganellus 15
Goby, Rock *15*
Godwit, Black-tailed 47
Goldcrest 55, 57
Goldfinch *49*, 60
Goshawk 32
Grebes 27, 28
Greenfinch 60
Greenshank 47
Grouse 35, 38
Grus grus 38, *38*
Guillemot 50
   Common *31*
Gulls 35, 42, 49
Gulo gulo 79, 92
Guppies 25
Gyps fulvus 32, *44*
Gyrfalcon 35

Haematopus ostralegus 42
Haliaeetus albicilla 32
Halichoerus grypus 99, 103, *105*

Hamsters 70, 74
Hare, Blue or Mountain *93*, 95
   Brown 79, 95
   European *93*
Hares 79, 81, 92
Harriers 34
Hawfinch *57*, 60
Hawk, Sparrow 32
Hedgehog, Algerian or Vagrant 67
   European or Spiny 67, *86*
Hemipode, Andalusian 35
Hen-fish *25*
Heron, Great White 29
   Grey 29, *36*
   Purple *36*
Herring 19, 24
Himantopus himantopus 47
Hippocampus antiquorum 22
Hirundo rustica 44, 52, 53
Hobby 35
Hoopoe 52
Hydroprogne tschegrava 50
Hyla arborea 123
Hyperoodon ampullatus 103

Ibex 90, 99
Ictalurus melas 16
   I. nebulosus 16

Jackdaw 61
Jays 61
Jynx torquilla 52

Kestrel 34, *42*
Kingfishers *40*, 52
Kites 32, *46*
Kittiwake 49, 51
Knot *31*

Labrus ossifagus 15
Lacerta agilis *110*, 111
   L. lepida *108*, 113
   L. muralis 111, *114*
   L. sicula 111
   L. viridis *108*, 111, *111*
   L. vivipara 110, *112*
Lagopus lagopus 35
   L. l. scoticus 38
Lanius collurio 54
   L. excubitor 55
   L. nubicus 54
Lapwing 42
Larks 53
Larus canus 35
   L. hyperboreus 49
   L. marinus 49
   L. ridibundus 49
Lemmings 94
   Norway 93, *96*
Lemmus lemmus 93, *96*
Lepus europaeus 79, *93*
   L. timidus *93*, 95
Leucaspius delineatus 19
Leuciscus leuciscus 16
Limosa limosa 47
Linnet 60
Lizards 107, 115, 116, 117, 121
Lizard, Common or Viviparous *108*, 110, *112*
   Eyed 113, *108*
   Green 111, *111*, 113, *108*
   Sand *110*, 111
   Starred Agama 114

Ruin 111
Wall 111, 114
Worm 110
Loach 20, 21
Lota lota 16
Lullula arborea 53
Lump-sucker *25*
Luscinia luscinia 59
   L. megarhynchos 49, *58*
Lutra lutra 81, *86*, *95*, *105*
Lynx, Northern 72, *83*, 93
Lyrurus tetrix 38

Mackerel, Atlantic *13*
Macroprotodon cucullatus 120
Magpie 61
Malpolon monspessulana 119
Mallard 30, *39*
Marmot, Alpine 88, *96*
   Bobac 87
Marmota bobak 87
   M. marmota 88, *96*
Marten, Beech or Stone 81
   Pine 81, *97*
Martes foina 81
   M. martes 81, *97*
Martin, House 53
   Sand *52*, 53
Megaptera novaeangliae 103
Melanitta nigra 31
Melanocorypha calandra 53
   M. yeltoniensis 53
Meles meles 81, *86*
Merganser, Red-breasted 32
Mergus serrator 32
Merops apiaster 52
Mesocricetus auratus 74
Mesoplodon bidens 103
Micromys minutus 71, *91*
Microtus nivalis 88
Milvus migrans 32, *46*
   M. milvus 32
Mink, European 82
Minnow 17, 18, *18*
Misgurnus fossilis 21
Moderlieschen 19
Moles 67, 70, *89*
Monticola saxatilis 58
   M. solitarius 58
Moorhen 39, *42*
Mosquito Fish 25
Motacilla alba 54
   M. cinerea 54
   M. flava 54
   M. f. feldegg 54
   M. f. flavissima 54
Mouflon 99
Mouse 70
   Harvest 71, 74, *90*
   House 71, 74
   Wood 74, *90*, *97*
Mud-minnow, European 23
Mugil spp. 15
Mullet, Grey *15*
Mus musculus 71
Muscardinus avellanarius 74, *97*
Muscicapa striata 57
Mustela erminea 81, *95*, *101*
   M. lutreola 82
   M. nivalis 81, *95*, *101*
   M. putorius 81
Myocastor coypus 75
Myotis mystacinus 84

Natrix maura 116
    N. natrix 107, *107*, 115, *118*
    N. n. astreptopha 116
    N. n. helvetica 115
    N. tesselata 116, *119*
Neomys fodiens 70
Neophron percnopterus 32, *46*
Netta rufina 31
Newt, Alpine *112*
Nightingale *49*, 58
    Thrush 59
Nightjars 52, *52*
Noemacheilus barbatulus 20
Nucifraga caryocatactes 61
Numenius arquata 39, *47*
Nutcracker 61
Nuthatches 59
Nutria 75
Nyctea scandiaca 51

Oenanthe hispanica 57
    O. leucura 57
    O. oenanthe 57
Ophisaurus apodus 110
Oriole, Golden 61
Oriolus oriolus 61
Oryctolagus cuniculus 63, 79, *81*, *93*
Osmerus eperlanus 24
Osprey 32, 34, *46*
Otis tarda 42
Otter, Eurasian 81, 86, *95*, *105*
Otus scops 51
Ouzel, Ring 59
Ovibus moschatus 64
Ovis musimon 99
Owl, Barn 50
    Eagle *50*, 51–52
    Great Grey 51
    Hawk 51
    Scops 51
    Snowy 51
    Tengmalm's *49*
Ox, Musk *64*
Oystercatcher 42

Pagophilus groenlandicus 103
Pandion haliaetus 32, *46*
Panurus biarmicus 59
Partridges 35, 38
Parus ater 59
    P. caeruleus 49, 59
    P. major 59
    P. palustris 59
Passer domesticus 60
    P. montanus 60, *61*
Pelecanus crispus 28
Pelecus cultratus 19
Pelicans 28
Perca fluviatilis 16, *18*
Perch 16, 17, *18*
Perdix perdix 38
Peregrine 35
Perisoreus infaustus 61
Pernis apivorus 32
Petrel, Storm 28
Phalacrocorax aristotelis 29
    P. carbo 28, *28*
Phalarope, Red-necked *35*, 47
Phalaropus lobatus *35*, 47
Phasianus colchicus 38, *44*
Pheasant 35, 38, *44*
Philomachus pugnax 47, *51*
Phoca vitulina *99*, 103

Phocaena phocoena 103
Phoenicopterus ruber 29, *32*
Phoenicurus ochruros 58
    P. phoenicurus 58, *60*
Phoxinus phoxinus 17, *18*
Phylloscopus collybita 56
    P. sibilatrix 56
    P. trochilis 56
Pica pica 61
Picus canus 53
    P. viridis 53
Pigeons 51, 55
Pike 16, *16*, 17, 22, 23
Pilchard 24
Pipefishes *13*
Pipits 53, 54
Plaice *23*
Platalea leucorodia 29
Plectrophenax nivalis 60
Pleuronectes platessa 23
Plovers 42, 46
Pluvialis apricaria 42
Pochard, Red-crested 31
Podiceps cristatus 27
    P. ruficollis 28
Polecat, European 81, 83
Pope *18*
Porpoise 103
Pratincole 47
Prunella collaris 55
    P. modularis 55
    P. montanella 55
Pterocles alchata 51
    P. orientalis 51
Puffin 32, *36*, 51
Puffinus puffinus 28
Pusa hispida 103
Pyrrhula pyrrhula 60

Rabbit, European *63*, 79, *81*, *93*
Rail, Water 39
Raja clavata 23
Rallus aquaticus 39
Rana esculenta 112, *123*
    R. ridibunda 123
    R. temporaria 107, *122*
Rangifer tarandus 71, 98
Rat, Black or Ship 74
    Brown or Sewer 74
    Lesser Mole 87
Rattus norvegicus 74
    R. rattus 74
Raven *46*, 61
Razorbill 51
Ray, Thornback *23*
Recurvirostra avosetta 47
Redpoll 60
Redshank 38
Redstarts 58, 59, *60*
Redwing 59
Reedling, Bearded *59*
Regulus ignicapillus 55
    R. regulus 55
Reindeer 71, 98
Remiz pendulinus 59
Rhodeus sericeus 20
Rhombus maximus 23
Riparia riparia 52, 53
Rissa tridactyla 49
Roach 16, 17, 18, *18*
Robin *49*, 58, 59, *90*
Roller 52
Rook 52, 61

Rudd 17
Ruff 47, *51*
Ruffe 16, *18*
Rupicapra rupicapra 89
Rutilus rutilus 16, *18*

Salamander *111*
Salamandra salamandra 111
Salmo gairdneri irideus 21, 23
    S. salar 13, 17
    S. trutta 17, *21*
Salmon *13*, 16, 17, 24
Samateria mollissima 31
Sanderling *31*
Sandgrouse 51
Sawbills 32
Saxicola rubetra 57, *60*
    S. torquata 57
Scardinius erythrophthalmus 17
Scheltopusik 110
Sciurus carolinensis 70, 77, 89
    S. vulgaris 70, *77*
Scolopax rusticola 46
Scomber scombrus 13
Scorpaena scrofa 15
Scorpion Fish, Orange *15*
Scorpion, Sea 25
Scoter, Common 31
Scyllium canicula 23
Sea-horse *22*
Seal, Common *99*, 103
    Grey *99*, 103, *105*
    Harp 103
    Ringed 103
Sepia officinalis 25
Shad 24
Shag 29
Shanny *13*
Shearwater, Manx 28
Shrews 70, *89*
Shrikes 54, *55*
Silurus aristotelis 22
    S. glanis 21
Sitta europaea 59
    S. neumayer 59
    S. whiteheadi 59
Skua, Arctic 47
    Great *42*
Skylark 53
Slowworm *108*, 110, *114*
Smelt 24
Snake, Aesculapian *116*, 117
    Barred Grass 115, 116
    Cat 119
    Cowl 120
    Dark Green 117
    Diced Water 116, *119*
    Four-lined 117
    Glass 110
    Grass 107, *107*, 110, 115, 116, *118*
    Hooded 120
    Leopard 117
    Montpellier 119
    Smooth 116, *119*
    Southern Smooth 120
    Viperine Water 116
    Whip 117
Snipe 46
Sorex araneus 70, 89
    minutus 70
Sousliks 86
Spalax leucodon 87
Sparrow, House 60

Tree 60, *61*
Spoonbill 29
Sprat 24
Squalius cephalus 16
Squirrel, Grey 70, 75, *77*, *89*
    Red 70, 71, *77*
Squirts, Sea *15*
Starling, Common *57*, 60
    Rose-coloured 60
    Spotless 61
Stercorarius parasiticus 42, 47
Sterna dougallii 50
    S. hirundo 49, *127*
    S. paradisea 50
Sticklebacks 16, *18*
Stilt, Black-winged 47
Stoat 81, 82, 83, *95*, *101*
Stonechat 57
Stork, White 29, *34*
Streptopelia decaocto 51
Strix nebulosa 51
Sturgeon 24, 25
Sturnus roseus 60
    S. unicolor 61
    S. vulgaris 57, 60
Sula bassana 28, *29*, *36*
Sus scrofa 64, *75*
Surnia ulula 51
Swallow *44*, 52, 53
Swans 29, *34*
Swifts 52
Sylvia atricapilla 55
    S. borin 55
    S. cantillans 56
    S. communis 55
    S. curruca 58
    S. rueppelli 56
    S. undata 56

Talpa caeca 70
    T. europaea 67, *89*
    T. romana 70
Tarentola mauritanica 115
Tarsiger cyanurus 59
Teal 31
Telescopus fallax 119
Tern, Arctic 50
    Black 50
    Caspian 50
    Common 49, *127*
    Marsh 50
    Roseate 50
Terrapins 122, 123
Testudo graeca 121
    T. hermanni 121, *121*
    T. horsefieldi 122
    T. marginata 121
Tetrao urogallus 38
Thalarctos maritimus 98, *99*, *103*
Thrush, Blue Rock 58
    Mistle 59
    Rock 58
    Song 58, 59
Tichodroma muraria 59
Tit, Blue *49*, 59
    Coal 59
    Great 59
    Long-tailed *58*, 59
    Marsh 59
    Penduline 59
Tortoise, European 121
    Greek 121, *121*
    Herman's 121

    Horsefield's 122
    Marginated 121
    Pond *121*, 122, 123
    Spur-thighed 121
Tree-frog, European Green *123*
Tringa nebularia 47
    T. totanus 38
Triturus alpestris 112
Troglodytes troglodytes 55
Trout 16, 17, *21*, 23, 24
Truncatus truncatus 105
Turbot 23
Turdus iliacus 59
    T. merula 58
    T. philomelos 58
    T. pilaris 59, *60*
    T. torquatus 59
    T. viscivorus 59
Turnix sylvatica 35
Turnstone *31*
Turtles 120
Tyto alba 50

Umbra krameri 23
Upupa epops 52
Uria aalge 31, 50
Ursus arctos 66, *75*

Valencia hispanica 25
Vanellus vanellus 42
Vimba vimba 19
Viper, Asp 117
    Long-nosed or Horned 118
    Northern 107, 116, 117
Vipera ammodytes 118
    V. aspis 117
    V. berus 107, *108*, *116*, 117, *119*
Vole 92, 116
    Bank *101*
    Snow 88
    Water *101*
Vulpes vulpes 67, 68, *79*, *81*, *82*
Vultures 32, *44*, *46*

Waders 42
Wagtails 54
Warblers 55, 56, *61*
Waxwing 54
Weasel 81, 82, *95*, *101*
Weather Fish 21
Wels 21, 22
Whales 103
Wheatears 57
Whinchat 57, *60*
Whitefishes 16, 17–18
Whitethroats 55, *58*
Whiting 19
Wigeon 31
Wisent 63, 64, *64*
Wolf *72*, *81*, 91
Wolf-fish *13*
Wolverine *79*, 92
Woodcock 46
Woodpeckers *44*, 52, 53, *55*
Wrasse, Cuckoo *15*
Wren 55
Wryneck 52

Yellowhammer 60

Zährte 19
Ziege 19
Zope 19

Photographic Acknowledgments
Heather Angel 17; John Bailee 29; S. Bayliss Smith 39 TL; Julius Behnke endpapers; S. C. Bisserot 84 L, 84 R–85, 110 B, 116 R–117, 119T; Jane Burton 14, 19, 21T, 24, 24–25 T, 68 B and front jacket, 76–77 L, 81 T, 100 L, 100 R, 106–107, 108 BL, 108 BR, 108 R–109, 112 T, 115; Arthur Christiansen 50 B, 93 TR; Werner Curth 40 L; Ken Denham 104–105; A. W. Engman 94; Gösta Håkansson 51 T; J. Hancock 48 B; R. A. Harris 67; E. Herbert, A.F.A. 3; P. Hinchcliffe 80; Ingmar Holmåsen 12–13 T, 22, 26–27, 28 L, 34-35 T, 49, 55, 71 BR, 79 B, 97 BL, 99 T, 113, 114 B, 118 TL, 118 B, 120–121, 121; Eric Hosking 31, 32 B, 35 TR, 35 B, 36 BR, 38 T, 39 B, 42 T, 42 B, 44 L, 46–47 B, 47 R, 50 T, 51 BL, 52 TL, 52 TR, 90 L, 90 R–91 L, 95 T, 96 T, 99 B, 99 C; Peter Jackson 36 TR; G. B. Kearey 92; Russ Kinne 12 BL, 39 TR; Geoffrey Kinns, A.F.A. 69, 73, 77T, 97 R, 105 T, 105 B; John Markham 16 T, 16 C, 18 T, 18 B, 20, 36 L, 45 BL, 46 T, 48 TC, 48 TR, 53, 56–57, 58 T, 58 B, 59 T, 59 B, 60 L, 60 C, 60 R, 61 C, 61 R, 74, 77 B, 89 R, 91 R, 93 TL, 93 B, 95 B, 97 T, 101 T, 101 B, 110 T, 111 B, 114 T, 118 R–119 L, 119 C, 119 B, 122, 123 T, 123 B; R. K. Murton 54 T; S. C. Porter 28 R, 57 R, 61 L, 88; W. Puchalski 38 B, 51 BR, 72 R, 83 T; Georg Quedens 98 L; Ann Schmitz 62–63; Toni Schneiders 34 L; H. Schrempp 48 TL; H. Schünemann 43; H. W. Silvester 32 T; Barrie Thomas 37; Ronald Thompson 12 TL; M. R. Tibbles 30–31 L; Tierbilder Okapia 54 B, 64, 66 T, 66 B, 71 T, 75, 79 T, 81 B, 83 B, 86 L, 86 R, 87, 89 L, 96 B, 98 R–99 L, 111 T, 116 L, 123 C; Tiofoto 65, 70–71 B, 78, 82, 102–103; Douglas P. Wilson 12 BR, 13 B, 15 TR, 15 CR, 15 BR, 21 B, 23 T, 23 CL, 23 CR, 23 B, 25 B; Mrs. M. A. Wilson 12 CR; Z.F.A. 33 and back jacket, 40–41, 44 TR, 44 BR, 45 R, 68 T, 108 T, 112 B; Zoological Society of London 72 L.
Map drawn by Denys Ovenden.

Endpapers: a family of Common Terns (Sterna hirundo).
Red Deer (Cervus elaphus).